Science, Technology, and the Economic Future

New York Academy
of Sciences

Science, Technology, and the Economic Future

Edited by

SUSAN U. RAYMOND, PH.D.

New York Academy
of Sciences

© 1998 by the New York Academy of Sciences

New York Academy of Sciences
Two East Sixty-third Street
New York, New York 10021
telephone: 212.838.0230, ext. 348
fax: 212.753.3479; e-mail: policy@nyas.org

The New York Academy of Sciences believes it has a responsibility to provide an open forum for discussion of current topics in science, engineering, and medicine, including issues of public policy. The positions taken by speakers at the Science Policy Association are their own and not necessarily those of the Academy. Moreover, the Academy, its Board, and its members have no intent to influence legislation by providing such forums.

ISBN 1-57331-147-2

Printed in the United States of America.

Contents

Preface

ver its 180 year history, the New York Academy of Sciences has often served as a bridge between scientific innovation and public policy.

To illuminate more clearly both the role of science and technology *in* public and private policy making, and the necessity of supportive public and private policy *for* science and technology, the Academy regularly invites both public and private leaders to address key issues at the intersection of science, technology, and societal progress. With innovation increasingly powering the global economy and touching virtually every aspect of every society, the imperative of informed policy is central.

The essays in this special volume reflect the perspectives of a variety of government, corporate, and university leaders. Technological change creates both challenges and opportunities for global economic competitiveness and for the continued enhancement of societal welfare. The challenges and the opportunities are complex; neither will be easily met. Leaders who have addressed the Academy have repeatedly emphasized the tangle of policy dilemmas, domestic and international, which confront a technology-intense world.

But we cannot shrink from the choices that technological advance poses for economies, societies, and, yes, the political process. The fact that policy decision-making will be untidy must not deter leaders in the science and technology community from actively engaging in and leading the public debate. Equally, decisions about the allocation of resources in the private sector will be of comparable, and perhaps greater, import. Private investment and innovation will shape future prosperity. Science and technology leaders must deeply understand the choices that private organizations face, and serve a ready resource to guide those decisions. Both perspectives—those of public policy and those of private initiative—are amply represented in this volume of essays.

We thank the many leaders who have participated in Academy forums and shared their insights with colleagues from a broad range of institutions and disciplines. The New York Academy of Sciences continues its commitment to building a forum in which science, technology, and public policy can join to serve society even more productively. ∎

RODNEY W. NICHOLS
President and CEO

SUSAN RAYMOND, PH.D.
Director, Policy Programs

Science, Technology, and the Economic Future

New York Academy of Sciences

I

Formulating Policy

"There is no such thing as a fixed policy," noted the eminent British diplomat Lord Salisbury, "because policy like all organic entities is always in the making." Science and innovation represent flows of knowledge. Their impact and implications are, therefore, often difficult to relate effectively to a stock of public policy. Knowledge changes constantly; policy changes only with great deliberation and much effort. Yet it is essential that public policy formulation ensure that the best scientific thinking and technological innovation is reflected *in* policy, and that policy *for* science and technology encourages dynamic flows of knowledge. This is an often difficult and controversial proposition. Policy formulation is both a careful and a political process. In turn, science and technology cannot escape things political when it seeks to affect policy nor when it seeks to be supported by policy. As the following essays illustrate, formulating effective and timely policies that have science and technology implications is a complex undertaking.

Steering National Security into the Next Decade: S&T as Navigational Device

Based on remarks delivered at the New York Academy of Sciences, June 25, 1996.

JANE WALES
Assistant to the President for International Security
The Rockefeller Brothers Fund

I am going to discuss the changed security policy agenda, the forces that shape it, the logic that drives it, and the mechanisms for tying science and technology to its aims. I will then say a word about the obstacles to success.

The world has changed, and with it the threats to and requirements of security. The bipolar conflict that defined the Cold War no longer dominates. And, at least three interacting and largely spontaneous trends now challenge the capacity of states to govern and of nature to provide.

The first is the communications revolution and the attendant diffusion of military and other technologies, of information and

> **We must rely on the promise of science and technology to expand both our knowledge and our options.**

ideas. Information technology has already had the effect of decentralizing decision-making and authority. It is fair to say we do not know the full consequences of this third Industrial Revolution. But we do know that it is powerful and transforming.

The second trend is the restructuring of the global economy; the redistribution of wealth, production, and power; and the exacerbation of income disparities within and among states. The combination of economic globalization and the information revolution has brought about a rapid shift in the structure of U.S. labor markets from manufacturing to knowledge-based jobs. And the income advantage of college graduates over high school graduates is widening quickly with real social consequences.

This would not look quite so unstable were it not for the third trend: population surges at the low end of the economy, adding almost a billion people per decade. This young, mobile population is moving to cities at the rate of a million per week, requiring water, sewage treatment, energy, transportation, and sources of employment. If fertility rates decline to replacement levels, we will have a world population of 10 billion in the year 2050, requiring a tripling of food and energy production over the next 50 years. We are not prepared.

Understanding these three forces and learning to manage the interactions among them have rightly become a preoccupation of scholars and policy-makers. We must rely on the promise of science and technology, not only to help us understand these forces, but also to contend with them—to expand both our knowledge and our options. Greater knowledge will help us to predict and prevent crises, to organize ourselves for decision-making and response, and to develop concepts of security that capture the synergies among the many contributors to conflict, so that we can treat them as symptoms of a syndrome, rather than separable, inexplicable, or unimportant events.

U.S. Government Organizational Innovations

The problems we face require cross-disciplinary analysis and cross-sectoral action. But we are not organized that way, either in government or in academia. So we have had to invent mechanisms for cooperation, among states,

> Demographic growth has created a young, mobile population that is moving to cities at the rate of a million people a week.

among government agencies, and among disciplines. I will point to three such innovations within the U.S. government: the National Science and Technology Council (NSTC), the program of comprehensive S&T cooperation known as "country strategies," and the creation of binational commissions led by the Vice President.

The NSTC is a Cabinet-level interagency body that is chaired by the President, managed by his Science Advisor, empowered to draft presidential directives, and mandated to set federal R&D budget and program priorities. Its members have hammered out interagency agreements on programs aimed at enhancing our security by advancing economic performance at home and sustainable development abroad. It has, for example, shaped investment initiatives in such areas as high-performance computing, environmental technologies, dual-use technologies, and the "clean car" initiative. It has developed integrated responses to specific problems, such as the threat of emerging and reemerging infectious diseases. An NSTC-crafted Presidential Directive establishes an international electronic network for surveillance and response

> **We must develop concepts of security that capture the synergies among the many contributors to conflict, so that we can treat them as symptoms of a syndrome, rather than separable, inexplicable, or unimportant events.**

to infectious diseases and establishes cooperation among state and local governments, international organizations, the pharmaceutical and other industries, the medical profession, and other nongovernmental players.

This initiative is supported by another NSTC activity—the development of strategies for comprehensive S&T cooperation with states that, by virtue of their size, geography, resources, economy, and politics have the capacity to either build or to undermine stability in the region. Countries such as China, India, Russia, South Africa, Argentina, and Brazil are poised to make decisions about how they will provide energy, transportation, communications, and food for their citizens. The development decisions they make can do irreparable environmental damage or reverse environmental decline, stabilizing both the physical and political environments. Each of these countries has a strong S&T community. This community is a force for political reform. Its participation in the economy is essential to economic reform. It will lead defense conversion efforts. It is frequently a voice for the protection of individual rights. And it represents the talent pool that will attract long-term trade and investment.

These "country strategies" have fed into and supported the third innovation that I will discuss—the establishment of bi-national commissions under the leadership of the Vice President. These commissions bring Cabinet-level attention to S&T cooperation aimed at addressing problems ranging from reversing "brain drain" from economies in transition, to ensuring cooperation in the pro-

tection and control of nuclear
weapons materials, to addressing
broad problems of public health
and food security, or to encouraging defense conversion by designing joint ventures. The Vice
President co-chairs the
Gore–Chernomyrdin
Commission with the Russian

> **We will not succeed in strengthening and preserving our S&T asset if we continue to act on a reflexive distrust of multilateral institutions and international cooperation.**

Prime Minister. He co-chairs the Gore–Mbeke Commission with
the South African Deputy President. And he has recently launched
a binational sustainable development initiative with the Chinese
Minister of State Science and Technology.

Just as in the worst of times with the Soviet Union we had a
mutual interest in arms control, so too in the worst of times with
China we have a mutual interest in sustainable economic development. But this is not obvious to all, especially in a time of renewed
isolationism and tremendous fiscal and political constraints. The
sustainable development initiative with China serves as a reminder
that security interests and scientific opportunities do not always
align themselves with domestic political concerns.

Science and Technology: A National Security Asset

I left government thoroughly persuaded that our nation's scientific and industrial base is its greatest security asset. But we will not
succeed in strengthening and preserving that asset if we continue
to favor short-term procurement over long-term investment; if
we act on a reflexive distrust of multilateral institutions and
international cooperation; or if we succumb to an ideology that
worships the market as the all-knowing distributor of burdens
and benefits and views governments not only as unable to solve
problems, but also as being their primary cause.

Let me close by saying simply that the NSTC, the country strategies, the binational commissions, and other mechanisms for S&T cooperation have served us well. But the intellectual and institutional frameworks for linking science, technology, stewardship, and security are in their early stages, are personality-driven, and could therefore be short-lived. ■

In Service to Society: Renewing the Contract between Science and Government

Based on remarks delivered at the New York Academy of Sciences, November 7, 1995. Professor Donald Stokes was a champion of science and technology. With his death in January of 1997, science has lost one of America's most astute observers of the importance of research and development to the nation's prosperity. The Academy and all of science mourn his passing.

DONALD E. STOKES

Professor of Politics and Public Affairs
Woodrow Wilson School at Princeton University

This is a period of extraordinarily rapid change, not least in federal science and technology policy. The changes in S&T policy have their roots in several background factors that are shaping the policy agenda.

Near-Term Factors in Changing S&T Policy

Certainly, one of the most significant forces has been the astonishing end of the Soviet challenge. The collapse of America's deadliest adversary has led government to reconsider the billions of budgetary dollars locked up in R&D accounts for defense. Although the debate over "how much" for defense R&D will continue for some time to come, the overall response, "less," is agreed on.

> **Budget deficits, and now the emphasis on reducing deficits, have left science and technology policy hostage to the budgetary imperative.**

The cauldron of global competition has been a second factor in changing thinking about R&D policy. As the U.S. economy becomes integrated into a more vibrant global economy, concern has shifted from technological confrontation with an adversary power to broader technological competitiveness in the marketplace. In that marketplace, the competitors are not military giants, but nations that are extremely technologically advanced. Hence, the imperative for federal policy has shifted to supporting innovation for rapid commercial, and hence economic, advance across the entire global stage.

The third factor within which S&T policy change is occurring is the legacy of past budgetary policies pursued by both political parties over decades in Congress and the White House. Budget deficits, and the current emphasis on reducing deficits, have left science and technology policy hostage to the budgetary imperative. The wisdom of existing funding relationships between science and technology and government is routinely measured against deficit-reduction goals.

The Deeper Background: The Bush Report

It would be a mistake to reconsider S&T policy issues merely on the basis of these near-term factors. Current events mask a deeper change in the canvas upon which science and technology policy has long been painted. In sum, there has been a gradual weakening of the conceptual foundations of the vision of basic science and its relationship to technological innovation that underpinned the great compact between science and government in the period after World War II.

Vannevar Bush's great report, *Science, the Endless Frontier,*[1] set out a conceptual and organization-al relationship between science and government that has held sway for over four decades. It is important to recognize that, in its conception, the report, and espe-cially the background panel led by

> **The Vannevar Bush report created a view of basic science and technological innovation that became the conceptual foundation of the post-war compact between science and government.**

Isaiah Bowman, represented a mechanism for achieving the essen-tially political objectives of the science community in the immedi-ate post-war period. The central objective was to extend the broad scope of the Office of Scientific Research and Development, which had managed science funding during the war, through the creation of a peacetime National Research Foundation. Optimally, that foundation would have been self-governing and, if the Bowman panel had its way, even exempt from the federal budget process because it would have had an expendable endowment that would need reprovisioning only at widely spaced intervals. The plan was to ensure long-term funding and remove the science-support process from the vagaries of near-term politics.

The plan, of course, was dead on arrival. The post-war political process was unreceptive to such grand schemes. Hence, when in 1950, five years after the Bush report was issued, the National Science Foundation was created, it enjoyed much less scope and was

---

[1]Vannevar Bush, *Science, the Endless Frontier,* National Science Foundation, Washington, DC, reprinted 1990. Bush was Franklin Roosevelt's director of the Office of Scientific Research and Development during World War II. His remarkable achievement was the recruitment of numerous gifted scientists for the wartime research effort, which culminated in the development of the atomic bomb. Late in 1944 Roosevelt asked Bush to look ahead to the role of science in peacetime. *Science, the Endless Frontier* is Bush's report, although Roosevelt had died before it was filed.

> **A one-dimensional graphic with two poles, basic research at one end and applied science at the other, became a popular illustration of the innovation process.**

required to answer to Congress for its budgetary allocations.

As the policy process shattered the original organizational conception, those who wanted to keep federal funding flowing and to drastically reduce the control of the federal government on the content of research turned back to the Bush report for inspiration. From that seminal document, they derived a view of basic science and technological innovation that became the conceptual foundation of the post-war compact between science and government.

Canon One: Research for Its Own Merits

That compact was impressed into two aphorisms that would be worthy of Francis Bacon. The first was that basic research, a term that Bush himself coined, is performed without thought of practical end. Although it sounds like a definition, that aphorism was not meant to be a definition. It was meant to express the view that, in basic research, there is a tension between the quest for fundamental understanding, on the one hand, and considerations of use on the other. By extension, there was a radical separation between the categories of basic research and of applied science. A one-dimensional graphic with two poles, basic research at one end and applied science at the other, became a popular illustration of this relationship. You cannot get closer to one of those polar opposites without getting farther away from the other. The point of the distinction was to convince the policy community that any attempt to constrain the free creativity of the basic scientist at the "research" end of the spectrum was self-defeating; constraint meant loss.

Canon Two: Basic Research Yields Technological Development

To provide ammunition for ensuring continued federal funding, a second canon was added to the first. Basic research leads technological development. If basic research is insulated from premature considerations of use, it will become a remote but powerful dynamo for technological innovation as the advances of basic science are converted into advances in technology by the process of technological transfer. Again the linear model—the technological prowess is achieved in steps from basic research to applied research to development to production and operations.

A Third Element: Capturing Returns

A third element of the post–World War II case for scientific funding does not have quite the conceptual standing of the previous two parts of the argument, but it was important to the case that Bush and his colleagues were trying to build. The belief was that a country that invests in basic science can expect to capture the technological return for itself. In turn, a country that looks to others for its basic science will be weak in industrial development and laggard in world trade. This articulation of the case soaked deeply into the consciousness of the policy community with the launching of Sputnik in 1957. If such nasty technological surprises were to come from the Soviet system, the argument went, then it must mean that the roots of the surprises were in the strength of Soviet science. On the heels of that argument came a surge of federal

> The linear view of science and technology clearly does not fit the complicated and unevenly paced reality of innovation. Science is not exogenous to technological innovation. Modern science deals with phenomena that are often revealed only by advances in technology.

funding for science, scientific advice directly placed in the White House, and the like.

But Is It True? A Change in Concepts

Does the Bush formulation of the science–technology–government relationship fit reality now? As with many simple conceptions, the answer is, not really.

Throughout the history of science, the link between basic research and application has been close and interactive. The rise of microbiology in the late 19th century is an absolutely classic example of the development of a fundamental area of science that was informed at every stage by considerations of use. The mature Pasteur never did a study that was not applied while he laid the foundations of modern microbiology. The Pasteur example is not isolated. Across the English Channel, the physics of Lord Kelvin was deeply industrial in its inspiration. On the Rhine, the organic chemists were achieving scientific wonders in order to lay the basis of the chemical dye industry that later gave birth to the pharmaceutical industry. In the United States, in the great years of the General Electric Laboratory, Irving Langmuir was fascinated by the surfaces of electronics components being produced by industry. In the several billion year history of the world, there had never been any analogues of those surfaces until they were presented by technology. It was technology that opened whole new horizons for basic research.

The linear, one-way flow view of science and technology that emerged from Bush's report, although never actually endorsed by him, clearly does not fit the complicated and unevenly paced reality of the relationships in the S&T innovation process. Science is not exogenous to technological innovation. Modern science deals with phenomena that are often revealed only by advances in technology. What is needed is not a linear model, but two trajectories, scientific understanding on the one hand and technological capacity on the other. These two trajectories are semi-autonomous. Science is advanced at times by

> **Neither the policy community nor the nation has lost its fundamental interest in enlisting the capacity of science to address the needs of society.**

pure research without any fresh intervention of technology. Technology is at times advanced by changes in design or tinkering at the bench without any fresh intervention from discoveries in science. But, at times, advances in each of those trajectories has massive impacts on the other, and the impacts can go in either direction. Use-inspired research has been enormously important in the experience of science at least since the late 19th century. That importance was concealed by the Bush framework as it was absorbed into the policy debate.

Having taken a more realistic view of the two canons of basic research that underpinned Bush's case, then, it is important also to examine the third element of the argument—that a nation investing in basic research can expect to capture technological improvement. Experience both before and after World War II leads to skepticism about that premise. Yankee ingenuity, as Bush himself pointed out, led this country to industrial leadership when its science was still far behind that of Europe. Japan has taught the world that lesson over again, as it established its technological leadership without a strong thrust in scientific research.

A great deal of basic science will be used by those who have the capacity to use it, regardless of where the science is carried out.

A New Conception—A New Compact

The United States needs a new conception of the process of scientific and technological development and, in turn, a new compact that relates S&T to government. There is growing recognition in society and in the policy community that basic science has not led to technological innovations that have met the full spectrum of society's needs. Indeed, some of those unmet needs have been created by technology itself. Many in the policy community are therefore responding: the Vannevar Bush deal is off. Open-ended government support for science has failed the test of producing significant social good.

Yet neither the policy community nor the nation has lost its fundamental interest in enlisting the capacity of science to address the needs of society. Indeed, the degree of public opinion support for science remains remarkably strong. A vision of science that acknowledges it as a resource that can be enlisted in service to needs has a better capacity to renew the compact between science and government than does the vision of science as pure inquiry without thought of practical ends.

With current concerns over budget deficits dominating the policy process, and with widespread concern about global competitiveness in both the legislative and executive branches of policy making, this new vision is essential to maintain government commitment to science. Only if the policy community becomes persuaded that there are real uses for the basic science being funded with public dollars, only if that community sees that science is really inspired by human need and not simply by intellectual curiosity, will leadership recommit itself to deep support for scientific funding. ■

The Policy Crisis Over Cryptography in the Information Age: Who Says Who Can Access What?

Based on remarks delivered at the New York Academy of Sciences, March 6, 1997.

KENNETH W. DAM

Max Pam Professor of American and Foreign Law
University of Chicago Law School

A s Americans we often pride ourselves on living in a free society. But we are now learning that our free and open society is a vulnerable society. Telecommunications and computer networks have increased vulnerability of both individuals and businesses, a vulnerability to which we voluntarily subject ourselves the moment we turn on the modem attached to our personal computers.

Individual and Corporate Vulnerability

As individuals, privacy is no longer an esoteric concern of intel-

> With the rise of telecommunications and computer networks, a free and open society is also a vulnerable society.

lectuals and civil libertarians. Health records, financial records, commercial transactions, telephone conversations—all are exposed to access, use, and abuse.

For businesses, electronic communications, confidential documents, business plans, and bids on contracts, are all subject to access through telecommunications. But it is not just private commerce that faces risk. National systems of air traffic control, energy, and transport are also vulnerable. John Deutsch, former director of the Central Intelligence Agency, once remarked that the electron is the ultimate precision-guided weapon. It can reach cleanly into any system, cutting through defenses, and accessing deep internal operations.

A central, necessary, but not sufficient defense against system vulnerability in an electronic age is cryptography, systems for scrambling data and messages. Cryptography operates at two levels. Confidentiality applications protect messages so that parties can not read a message even if they intercept it. Authentication bars third parties from even accessing a system unless they can authenticate themselves as intended users of the system. So cryptography is an essential part of the solution, but therein also rests a critical problem, because some government agencies believe that cryptography itself is the problem.

The Department of Justice and the Federal Bureau of Investigation (FBI) see cryptography as protecting crime syndicates and terrorists. If such groups use encryption, and if the authorities cannot read intercepted or wiretapped messages, or at least cannot read them in real time, it is feared that protecting the public or the national interest will be difficult. This tension between privacy, security, and law enforcement underpins much of the policy debate.

The Voyage of the Clipper Chip

It was against this background that the U.S. government made its ill-fated Clipper Chip propos-

> Cryptography is an essential part of the solution to the problem of vulnerability. Some government agencies, however, think cryptography itself is the problem.

al. To over-simplify a bit, the government proposed that every citizen's telephone, fax machine, and computer—business and personal—that was hooked up to a modem be equipped with an encryption chip. On the positive side, this would protect all such systems from hackers and crooks. The keys to the encryption would be deposited with government agencies so that the FBI and local law enforcement officers, under a properly entered court order allowing a wiretap, could listen in to personal and business communications. This concept, is known as "key escrow"; the keys to the encryption are held secret until and unless a court order allows their use.

An explosion of protest ensued. Not only did some consider the court-order process inadequate, there was concern that the very existence of the keys would allow access to private communications without court orders. The debate tapped a deep distrust of government as well as concerns over privacy. Although the original proposal was modified many times, controversy remains. On October 1, 1996, the government's present approach was announced. This approach would make any domestic "keys" system absolutely voluntary. Encryption would be based on software, not on hardware chips, even though it is clear that software is more vulnerable to tampering than hardware. Moreover, key-holding responsibility would be split among more than two private holders known as "escrow agents" or "trusted third parties." The FBI or other law enforcement authorities could still obtain the keys with proper court orders. Additionally, a qualified corporation could hold its own keys if it established a system whereby targeted employees would not learn when a valid court order arrived, thereby permitting surreptitious wiretapping.

Nevertheless, vulnerabilities would still exist with a private system. If some third party penetrated the key escrow agent, they would have access to the keys of everyone using the agent. It would be an "open sesame" nightmare. Key escrow agents must operate through human beings, and where humans are involved, bribery, collusion, and coercion is always a worry.

> **Present day export controls prohibit export of software or hardware capable of encrypting above 40-bit level. But graduate students can now crack 40-bit encrypted messages in less than four hours. The effect on U.S. software exports is potentially chilling.**

Implications for Export Controls

Among the most important implications of the October 1996 recommendations were their effects on exports. With certain exceptions, current export controls prohibit export of software or hardware capable of encrypting above a so-called "40-bit" level. This level is known as "weak cryptography"; Berkeley graduate students have cracked 40-bit encryption in less than four hours. With a powerful, dedicated mainframe computer, with software optimized for the task, the time needed to overcome such encryption is essentially zero.

In addition to the debate over domestic vulnerability, the U.S. software community has obvious market concerns. Under 40-bit restrictions, U.S. companies would lose their dominant position in pre-packaged software. If they cannot offer strong cryptography in foreign markets, their competitors will. In response, the Government would allow U.S. software firms to sell 56-bit software (which is sixty-five thousand times more secure than 40-bits) provided they entered into an agreement with the U.S.

> **There is no clear answer to the crisis. Law enforcement and national security concerns do conflict with individual privacy, with the legitimate needs of business, and with the international competitiveness of the software industry**

Government to have a system, as of two years from January 1, 1997, which automatically escrows keys. Key escrow would then be a condition of export.

Several firms have entered into these government agreements, but many in the industry are still extremely dissatisfied. In fact, no one is satisfied, least of all the FBI, which sees a very real threat to protection against terrorism. So far, Congress supports loosening export controls on encryption and allowing 56-bit encrypted software to be exported. If we lose the World Trade Center, or even a mid-sized building to a terrorist attack where failed wiretaps in the face of encryption can be blamed, however, congressional reversal is certain.

There is no clear answer to the crisis. Drug lords and terrorists can and do use cryptography. Law enforcement does rely heavily on wiretaps in response. Law enforcement and national security concerns do conflict with individual privacy, with the legitimate needs of business, and with the international competitiveness of the software industry.

Enter the National Research Council
Given all of these considerations, Congress mandated the

National Research Council to develop a report on the dilemma. The report, developed by a committee that reflected the views of business, law enforcement, software designers and other diverse interests, made a series of unanimous recommendations. It was this report that, according to the Administration, provided an impetus to the October 1996 recommendation for liberalization of export controls.

Crime in the streets may be in decline, but crime in the suites is on the rise.

The report reached several fundamental conclusions. First, computer and telecommunications security is still undervalued in American society. There is no central focus of responsibility for these issues. The National Security Agency is responsible for computer security of classified material. The National Institute of Science and Technology (NIST) of the Department of Commerce is responsible for sensitive, unclassified government-information traffic. The Federal Reserve Bank has some responsibility for banking. The list could go on. The point is that no one in the government has responsibility for a government-wide view.

Second, the business of law enforcement is to reduce the crime rate. Computer-based crime rates are rising. These crimes tend not to be as high-profile as violent crime, since victims, especially businesses, are motivated to keep the crimes quiet. While crime in the streets may be in decline, crime in the suites is on the rise. Yet electronic commerce has the potential to usher in the next industrial revolution. It cannot achieve its full potential without resolving the security problem. This will require two initiatives. First, the Committee recommended the liberalization of export controls in order to address the problem of vulnerabilities abroad and the position of the U.S. software industry in international markets. It also recommended that, although key escrow systems have many flaws, they should be tested to determine if a workable approach could be developed that would ensure wider security and yet meet law enforcement needs.

Third, insofar as national security is concerned, the most important foundation of security in the future will be the strength of the U.S. economy. The vulnerabilities of U.S. corporations, particularly abroad, is of greater concern than the proliferation of high-quality cryptography abroad. Such proliferation will happen, whatever actions the U.S. takes and whether or not it exports at all. Hence, the focus should be on the security needs of the U.S. economy. Recognizing the necessity of encryption, then, from a national security viewpoint, the nation is better off with a strong U.S. software industry abroad, essentially dominating world markets.

Domestic Action and International Reaction

But these issues obviously have spread beyond U.S. policy alone. Some U.S. trading partners have made it clear that they will impose import controls to counter any liberalization of U.S. export controls, largely out of fear of the deleterious effect on anti-terrorism strategies. Some countries do not know which side to support because they have as many internal economic and political interests at stake and in conflict as in the United States. The U.S. has now appointed an ambassador at large, a "crypto-envoy" if you will, David Aaron, whose job is to work with other governments in coordinating policy.

International solutions are critical. Papered-over diplomatic language and compromises will not suffice. Interoperable communications systems must work internationally if the promise of globalization is to be realized in economic terms. Yet domestic interests and concerns are real in nearly every nation. Control over access to crypto keys, together with law enforcement priorities and issues of economic competitiveness, create a complex set of conditions of engagement in what might be called the opening skirmish of the crypto wars. ■

To Provide for the Common Defense: Issues in Technology and Defense Strategy for Future National Security

Based on remarks delivered at the New York Academy of Sciences, February 5, 1997.

JAMES A. THOMSON

President and Chief Executive Officer, RAND

With the end of the Cold War, there has been considerable discussion of the opportunity for shifting defense resources to domestic programs. The process of that change, however, may be longer than most observers anticipate due to the pacing of technological change. The Department of Defense now faces a need to plan for replacing those systems that were purchased in the 1970s and 1980s, systems which are now coming to the end of their technological lives. Looking out to the year 2010, there will be a tremendous procurement need even with a smaller military force. The key question is, how do we think about such decisions? Given that there is no vital U.S. national security interest directly and immediately threatened on earth today, on what basis do we plan?

Getting the Definitions Right

First, clarification of definitions is important. When I use the term "national interest," I refer to interests beyond our shores. This is not a matter of domestic tranquility or safety in the streets. Moreover, the term "vital" also needs to be clarified. Politicians seem to find it necessary to attach the adjective "vital" to everything in order to justify their actions. But the word has a precise meaning.

The Commission on National Interest recognized that there are a hierarchy of national interests, with "vital" being the most important. The vital interest of the United States of America is to continue to exist and to maintain intact the political institutions that protect the freedoms and lives of its citizens. These are the interests for which the nation would be willing to spill American blood and spend vast amounts of its treasure, even if no one else in the world thought it was a good idea and was willing to join the cause.

> No vital U.S. national security interest is directly and immediately threatened anywhere on earth today, despite the fact that politicians find it necessary to attach the adjective "vital" to practically everything in order to justify their actions.

This is a tall order. We would act no matter what others thought, and secure these interests above all others.

Using such a definition, what are America's vital interests? The Commission identified five:

- to prevent, deter, and reduce the threat of nuclear, biological, and chemical attacks on the United States;

- to prevent the emergence of a hostile dominating power or group of powers abroad;

- to prevent the emergence of a hostile power on U.S. borders or in control of the seas nearby;

- to prevent the catastrophic collapse of major global systems, including trade, financial markets, supplies of energy, the environment; and

- to ensure the survival of U.S. allies.

The last of these five is debatable as a vital national interest. As Lord Palmerston remarked, nations have no permanent allies, only permanent interests. Classifying allies as an interest confuses means and ends. Alliances are not ends, they are means that we use to secure our interests.

Are Vital Interests Now Threatened?

What is the situation today with regard to vital national interests? Although we face potential threats, there is no adversary that has the capability and the intention (that is, the predisposition) in a potential crisis to strike the

> **What interest is "vital?" That for which we would be willing to spill American blood or spend vast amounts of our treasure, even if no one else in the world thought it was a good idea.**

> With respect to the major global systems, the one that is clearly of most concern is the energy supply from the Gulf, and the dependence of Western nations on that energy.

United States with nuclear weapons. The Soviet Union has de-targeted its nuclear weapons. The Chinese have the capability to launch attacks to strike American soil, but the relationship between China and the United States is such that this is not a current threat. Other countries do not have a combination of capability and desire.

Terrorism, of course, always lies in wait. But terrorism requires both national security and domestic planning. Similarly, there appear to be no nascent hegemonic powers on the scene. Economics impede such desires on the part of most nations. As for American borders, there is not even a sign of a potential threat. With respect to global systems, clearly the one of most concern is the energy supply from the Gulf, with a dependence by the West that is likely to spread to Central Asia. In the near term, concerns revolve around Iraq and, to a lesser extent, Iran. Over time, Russia and China also may pose dilemmas. But these are extremely hypothetical and long-term scenarios.

From the point of view of current defense planning, the United States is very secure. So, again, if there is no direct threat, how do we plan for national defense?

Planning for Defense When There Is No Threat
The initial approach, developed in the Bush Administration and taken up by the Clinton Administration, is called Major Regional Contingencies. Defense plans and programs would be constructed around two nearly simultaneous threats, one over oil in the

Gulf and a second in Korea. The advantage of this approach is that it is clear and understandable to traditional threat-based planners. Second, the scenario was fairly stressing: the military would have to plan to mount a major response in far away loca-

> **The United States must constantly strive to ensure that its technology is at the cutting-edge, because we have learned from history that technology is a deterrent.**

tions, and on a rapid timeframe, against foes with not insignificant military capabilities.

The disadvantages of this approach are also clear. First, the scenarios are incomplete in that they do not match the current distribution of American forces worldwide. There's no obvious explanation as to why we have forces in Europe. Second, and most importantly, these scenarios end when the particular threats disappear. If North Korea and/or Iraq collapses, what then for defense?

A more complex approach to threat-based planning would have created a whole portfolio of scenarios against which we would test capabilities. The problem with that approach is that the scenarios, if ever made public, would create foreign relations headaches between the U.S. and the hypothetical scenario foes (e.g., Brazil, Russia, China, and the like). Leaving an ally with the impression that you are planning for a contingency in which they would be an adversary does not promote smooth diplomatic relations. It is very difficult to place your foreign policy and defense planning on the same basis because of the possible negative interaction between the two.

An Alternative: Deterrence-Based Defense Planning

Is there a broad concept instead of major regional contingencies to guide our defense planning? Recalling the earlier definition of "vital national interest" and the current state of security, I believe that the highest goal of foreign policy and defense policy is to keep it that way—no direct and immediate threat to vital interests. The defense planning basis, then, is not threat-reaction, but deterrence. U.S. policy needs to focus on shaping policy environments around the world. There are two immediate consequences of thinking in these terms.

> **Military affairs are being changed fundamentally by technology.**

First, the presence of American troops overseas is essential. Second, the U.S. should constantly strive to ensure that American technology is at the cutting edge. History teaches that technology has an important political effect; it deters initiatives from potential foes at both the military and the political level. Recognition of a rival's technological superiority leads one to negotiate not to attack. Strategies for deterrence should also be applied to the approach to proliferators; the will and capacity to act could act as a deterrent should any proliferator seek to evolve into a threat to the United States.

A third issue raised by a deterrence approach to planning is the revolution in military affairs. The military is being changed fundamentally by technology, particularly information technology. The ability to use technology to strike from a great distance leads to quite different notions of what military forces should look like, especially on the ground. There are those who would bet now on new technologies and fundamentally change weapons and military organization using the next generation of procurement resources. But caution is in order. When I first began in military analysis in 1974, a new generation of technology was thought to be on the horizon. Some of those "new" technologies are still not in the field, nearly a quarter century later. Military forces and organization are not simply a matter of technology. They are also a matter of adapting technology to the military environment. That is often neither simple nor easy.

Changes, NATO, and Allies

Finally, two related issues face military planning. First, what will be the effect of NATO expansion on military organization and deployment? If the U.S. military, through NATO, takes on the role of maintaining ground force capability in Europe to defend countries east of Germany, the issue is not necessarily one of U.S. force size. More troops may not be needed. But such an expanded NATO does imply that those forces present in Europe are strategically nailed down; they cannot be used elsewhere. Today, those forces are expected to be available for conflicts elsewhere, principally in the Gulf. If expansion means they cannot fulfill that role, then the U.S. needs a bigger Army. This is a very serious planning, and political, issue which virtually no one wants to discuss. It is, however, very real.

Second, post-Cold War, what is the role of U.S. allies. Defense planning over the last forty years has assumed that America's allies would help defend joint interests. However, our allies have cut their defense budgets much more deeply than America has. They have sharply reduced procurement and R&D has dropped equally sharply. The effect is beginning to be felt throughout European forces. The U.S. is now in a situation in which our allies may in fact be technologically obsolescent. If a military action is required, the U.S. may not want its allies to assist because they could be more of a problem than a help.

Obviously, that is not an acceptable solution. But the status and prospects for allied technological capability is a serious issue, and a concern which must be addressed in both military and foreign policy hallways. ■

II

Assuring the Environment

No policy topic is more current, and often controversial, than the state of the environment. Amidst the turbulent public and political waters that often accompany environmental debates, the deeper, quieter, stream of scientific understanding is often lost. Equally, debate seldom considers the future in terms of the implications of knowledge and innovation for the ability to manage current environmental problems. In part, this is because careful deliberation and long-term thinking are always lost in the hurly-burly of public debate. However, it may also be because the scientific community itself is not organized to inject knowledge and innovation more forcefully into matters of public policy. The commentaries that follow both address the need for a stronger scientific voice, and provide illustrations of the importance of careful analysis and innovation as an underpinning for environmental policy.

Protecting the U.S. Environment in the Wake of Regulatory Reform: The Importance of Science

Based on remarks delivered to the New York Academy of Sciences on November 14, 1995.

G. Jon Roush

Senior Fellow, The Conservation Fund

A mericans have disagreed, and will no doubt continue to disagree, about the specifics of environmental policy. Under the twin banners of regulatory reform and deficit reduction, these disagreements have produced an attack on environmental laws. The preferences of the American people, however, expressed through recent elections and opinion polls, have re-emphasized the degree to which environmental protection remains a priority for the nation.

The Assault on Science

Beyond any particular law or environmental issue, we have a greater cause for concern regarding environmental policy: increasingly, environmental lawmaking seems disengaged from the base of environmental science. Institutions that historically have served as the reservoirs of scientific findings for making and evaluating laws are now threatened. Some have been eliminated. Perhaps the most notable instance is the demise of the Office of Technology Assessment in 1995. With the passing of the OTA, Congress lost its main source of information about science and technology. Nothing has taken its place, and lawmaking is the worse for the loss.

I would be the first to agree that many environmental policies require regular review and reform. Laws such as the Endangered Species Act, which was the source of so much acrimony in 1995, need to reflect the changing needs of the people (for example, small land owners) as well as the changing base of scientific knowledge. But the answer is not wholesale abandonment of policy. We need reform based on careful observation (for example, we still lack basic information about the location and condition of the elements of biodiversity in America) and rigorous experimentation with alternatives. We are destroying the institutional capacity to carry out the science that will enable wise reform, with results more serious than simply endangering particular laws. We are cutting into the very muscle and sinew of knowledge upon which wise policy depends, today and into the future.

Deeper Implications for Resource Management

The deep suspicion of science, and the rise of "junk science" in policy making, would be laughable if it weren't so tragically harmful. America is enduring fundamental assaults on the generation and use of knowledge as a basis for resource management, at least when knowledge might contradict other, more powerful agendas and special interests. The assault is not always blatant; it can be quite subtle. Blatant or subtle, the effect is the same—to discredit the utility of science as a foundation for resource management decisions.

The debate over cost-benefit analysis is a case in point. Congress calls for rigorous cost-benefit analysis to justify regulatory initiatives. Usually the cost of regulatory action can be calculated with fair precision. That is true across a range of environmental issues, from the cost of abating pollution to the cost of complying with ozone requirements to the cost of leaving wetlands alone rather than filling them for parking lots. Yet unlike the costs, the benefits are often difficult or impossible to calculate. How can we quantify the benefits of human life or the quality of human life? How can we quantify the benefits of an environment full of life?

> Increasingly, environmental lawmaking seems disengaged from the base of environmental science.

To confess our inability to quantify these things is decidedly not to say that they are valueless; it is to acknowledge that their evaluation involves ethics as well as information. To pretend that science alone can quantify these benefits is a perverse interpretation of science, a perversion born either of ignorance or hypocrisy. To call for cost benefit analysis as the sole decision-making tool, and then to fail to acknowledge its limitations and build that reality into decision-making, is a subtle attack on science itself.

Three Invisible Decisions Facing the Nation

These are not partisan issues. They are simply issues of congressional decision-making. Environmental wisdom and, equally, environmental foolhardiness are the preserve of no one party nor of any single branch of government. Conservation, good government, and good science are nonpartisan endeavors. There are friends and enemies on both sides of the aisle, although in the last few years the Executive Branch has shown itself to be a more consistent friend of the environment overall than has Congress.

As a nation, we are making three especially important but invisi-

> We are cutting into the very muscle and sinew of knowledge upon which wise policy depends, today and into the future.

ble decisions. They are invisible because they will not emanate from a single rational debate nor from a single group of people acting in one place at one time. They will emerge from incremental, sometimes imperceptible decisions and actions by all Americans, which in the aggregate will determine our environmental directions.

First, how will we manage large, complex ecosystems? Many environmental systems are immensely complicated, chaotic, and poorly understood. Such systems have biological components, but also humanly created systems of economic, political, and social dimensions. For example, how shall we coordinate the actions resulting from multiple human decisions within a single, large watershed? The principles that we choose for ecosystem management will, by and large, determine the degree of success we have in passing a sustainable world on to our grandchildren.

Second, what should be the role of scientific thought and information in the formulation of policy? Environmental decisions must have a scientific base, but choices about the environment are also premised on values. How do we want science and values to interact? What institutions will inject science into policy, or keep it out? To what extent do we want to protect science from politics, or should we?

Third, what values should guide our management and use of public resources? Indeed, this question is at the heart of many of the nation's most contentious environmental issues. It is a question we must face with agonizing urgency. What values should guide the management of public land and water? Should we privatize or subsidize public land? How far should we go to conserve our common stock of biological diversity? How will we resolve the tensions between the private and the public?

The Private and the Public: Finding the Balance

The third invisible decision drives to the heart of a central dilemma. From the beginning, America has lived with a tension between the values of private, individual enterprise and the values of a community in which people share a responsibility.

> The public do not need to be experts, but they do need to understand concepts such as gap analysis, economic development, and social impact assessment.

By and large, the tension has been healthy; it has given us resilience. The question now is one of balance. How much community are we willing to sacrifice for private gain?

Across the country, a wide variety of experiments are under way involving both private stakeholders and public agencies in ecosystem management. They are revealing a great deal about what works and what does not, as well as much about America's values and preferences. For example, in 1994, The Wilderness Society sponsored a one-year study with the University of Michigan to examine locally or regionally organized ecosystem management projects and to identify elements necessary for success. Four key findings resulted. First, successful ecosystem management needs expansive, and often expensive, technology. Information systems are particularly important. Second, success depends on good scientific information. Project leaders interviewed during the study placed utmost importance on reliable data. As one expressed it, "Use the best science available, and if it is not available, go out and get it." Third, both public officials and citizen participants need a high level of education about ecological and social systems. The public do not need to be experts, but they do need to understand concepts such as gap analysis, economic development, and social impact assessment. Scientists need to reach out to the people. Finally, most projects in the study benefited from public investment. Government

> **I would call on scientists to accept the responsibility that comes with expertise, to become scientist-activists.**

usually provided direct funding or tax credits, information, or technical assistance. Even people who initiated projects with an anti-government bias tended to modify that bias as they gained experience.

So, successful project managers cite good science, good information, and government assistance as critical factors of success.

A Call to Arms for the Scientific Community

The combination of these two trends—the attack on science as the underpinning for environmental policy and the expressed need for good science by ecosystem managers—represents a call to arms for the scientific community. If scientists sit on the sidelines, much will be lost. What can scientists do? I would call on them to accept the responsibility that comes with expertise, to become scientist-activists. I don't ask that all scientists agree with my views about policies; I ask that all agree about the importance of information and science. Step out of the laboratory on occasion and enter the political debate to correct errors and defend informed inquiry.

What can scientist-activists do? Offer testimony, write letters, call talk shows. The air is filled with inaccuracies. Every time those inaccuracies go unanswered, it is a victory for anti-scientism.

Scientists also need to defend the institutions that mediate between science and policy. Respond to attacks on such organizations as the National Biological Service or the National Institute for Science and Technology. The real act of destruction in the attack on science is to drive a wedge between rational people. If that wedge deepens, the future will not be business as usual or science as usual. The scientific community must step in to defend science and, in so doing, defend democracy. ■

First Forum on Science and Technology Goals: The Environment

Based on remarks delivered to the New York Academy of Sciences on September 18, 1996.

JOHN F. AHEARNE

Director, Sigma Xi

F inding a way to develop a national consensus on something as complex and potentially divisive as environmental priorities for the nation is a difficult, yet critically important problem. A concerted and unique effort to gather and combine a comprehensive set of views and opinions on future environmental priorities, National Forum on Science and Technology Goals, took place under the auspices of the National Academy of Sciences in 1995. It is important to understand the National Forum process to understand the importance of the findings.

The National Forum Process

In 1992, the Carnegie Commission on Science, Technology, and Government recommended that a series of national forums be convened in which a broad array of opinion would be sought to reach

> **The population growth problem is extremely sensitive, with all of the complexity inherent in the human condition.**

consensus on the potential contributions of science and technology to meeting long-term societal goals in the United States. With the support of the Carnegie Corporation of New York, the National Academy of Sciences organized a National Forum in 1995. Environment was chosen as the first topic. The Forum began its work by seeking the views of a wide range of individuals and organizations on how science and technology could address problems of the environment. On the basis of the framework that emerged from that survey, the National Academy commissioned nine technical papers and then invited 68 experts to four days of discussions in plenary sessions and small groups to generate ideas and to discuss priorities. Both the original survey and the subsequent conference attendees were selected to mirror the variety of views and institutions in the nation. The process involved federal and state government officials, industry, nongovernmental organizations, and activist groups. The Forum sought, and received, a broad spectrum of opinions.

The result was extremely broad ranging. The organizing committee took this range of views and developed six critical areas of focus that are reported in the final National Research Council publication.[1]

Economics and Risk Assessment

One of the most important conclusions in the report, and one with which some technologists might disagree, was that social science has a substantial contribution to make to environmental issues, and one of its particular assets is economic analysis and risk assessment.

[1] *Linking Science and Technology to Society's Environmental Goals,* National Academy Press, Washington, D.C., 1996.

Much can be contributed through this analytic technique, and much improvement in the technique is possible.

Risk assessment and cost-benefit analysis can be substantial aids to environmental decision-making. Nevertheless, improvements are needed. Present regulatory strategies frequently do not differentiate sufficiently between minor and major risks. The costs incurred to reduce risks often do not bear a consistent relationship to the magnitude of the risk involved and the number of people potentially affected. More research is needed to improve the analytic tools available to decision-makers. New approaches are needed for assessing the value people place on the services provided by ecosystems: Can models better accommodate such values? Can they predict changes in human behavior? Can quantitative risk assessment and cost-benefit analysis be integrated so that the health and ecological factors studied are the ones that the public understands and cares about?

> Social science has a substantial contribution to make to environmental issues, and one of its particular assets is economic analysis and risk assessment.

New economic strategies, using the incentive approach rather than the command and control approach, have significant advantages in the long run, both in making environmental improvements more effective and in reducing the resources needed to accomplish that end. However, incentive-based approaches have proved harder to apply than theorists had predicted, although the results are in agreement with predictions. More experimentation is needed.

Environmental Monitoring in Ecology

Monitoring is critical to better understanding of ecological systems. But the data available to predict trends in the environment

> **New economic strategies, using the incentive approach rather than the command and control approach, have significant advantages in the long run, both in making environmental improvements more effective and in reducing the resources needed to accomplish that end.**

are decidedly poor. Few measurements are taken over extended periods of time in the same manner or in different places that are directly comparable, the Hawaiian CO_2 data being the most notable exception. The desirability of directly comparable measurements has long been recognized, but the great amount of environmental data collected over the past several decades has generally failed to meet these standards. Indeed, there is often a general lack of agreement on the basic characteristics to be measured. In short, the U.S. monitoring system is outdated and inadequate.

A pressing need exists for compiling data that will characterize environmental change, evaluate the consequences of human activities, and provide an information base for sustainable management. The Office of Science and Technology Policy in the White House should review and evaluate the quality of existing measurement and monitoring systems. For its part, the Congress should assign either an existing or a new federal research organization the mission of working with the scientific community to identify the key subjects for ecological research, aimed at identifying and developing reliable indicators of the health and sustainability of the environment in ecosystems and establishing monitoring systems that meet society's decision-making needs.

Chemicals in the Environment

Of particular importance are new chemicals in the environment that are a source of stress on the ecosystem. This includes chemicals that have unanticipated by-product effects and chemicals whose daughter products from degradation can be dangerous. The pollution threats are to both air and water. Whereas 25 years of progress in air and water quality systems is to be applauded, there is still a great need to improve the ability to predict the environmental consequences of

> **Energy is critical to economic progress not only in rich nations, but, even more so, in nations seeking to climb farther and more rapidly up the ladder of global prosperity.**

a new chemical on a variety of scales before the great expenditure of resources involved in getting that chemical to the market is undertaken.

Our predictive modeling abilities are poor, especially for the chronic effects of chemicals in the environment. Better test methods are needed to evaluate models and monitor the long-term environmental impacts of single compounds that are emitted as a result of new products or processes. The models must extend to both by-products and degradation products of such products or processes, and to the biochemistry of both plant and animal species. Standardization of testing should be international, as should responsibility for testing.

Energy Systems

Energy is critical to economic progress not only in rich nations, but, even more so, in nations seeking to climb farther and more rapidly up the ladder of global prosperity.

Yet, the environmental effects of meeting these energy needs, particularly by the use of fossil fuels, can be large. The impacts are local, regional, and global. The most environmentally troublesome aspect of the present global energy system is the use of fossil fuels. Worldwide defossilization, however, would require

> **The population growth problem is extremely sensitive, with all of the complexity inherent in the human condition.**

much greater emphasis on three major R&D directions—renewable energy sources; energy efficiency and conservation; and safe, publicly acceptable nuclear power.

Sustained research will lead to more options for energy generation and use, less emission of carbon to the atmosphere, and more efficient use of natural resources. For electrical power, more research is needed for nonfossil fuel sources because per capita electricity use is strongly correlated with development. Research efforts are necessary for renewables, such as photoelectric cells and biomass, whose widespread use will not occur until the cost of electricity from these sources is so low that large public subsidies are no longer required to make them cost competitive.

Use of coal is another critical research target. Sixty percent of the electricity generated in the United States comes from coal. China, Russia, and India likely will rely to a large extent on coal. Improving the efficiency and reducing the emissions of coal plants will benefit the United States directly, and also the globe through the impacts of improvements in Europe and Asia.

Similarly, nuclear energy requires research attention. The nuclear industry has been crippled by the high cost of nuclear plants, government inability to solve problems of the safe and reliable disposal of waste, and the subsequent disenchantment of investors and the public. Research is required in all of these areas.

Another key area for attention is transportation. More than 50% of energy from petroleum products is used in transportation. Improving the fuel efficiency of automobiles will help with this problem, but another important strategy is switching from vehicles powered by internal combustion engines to electric or hybrid electric or hydrogen-fueled cars.

Industrial Ecology

Industrial ecology refers to the analysis of how the entire process of manufacturing can be improved by focusing on efficiency and environmental impact, not only within one company or industry, but also across industries. The objective is to develop strategies to integrate the design, production, and consumption of products to reduce the use of resources. One industry's wastes become another industry's input. Reusing and recycling "waste" may open up new opportunities for efficiency.

Currently, the approaches to industrial ecology are experimental and not completely agreed upon or accepted in many industries. Indeed, the increasing decentralization of formerly vertically integrated industries presents new problems for implementation of industrial ecology methodology. Nevertheless, the concept has intrinsic merit, and a great deal of further research and experimentation is called for.

Population

Global population trends bode ill for the environment. As noted in the formal National Forum report, over the next 50 years, global economic output is expected to quadruple and total population to double to about 11 billion people. Such population growth, in the context of existing modes of industrial and agricultural production and consumption, has the potential to significantly impact the environment.

The population growth problem is extremely sensitive, with all of the complexity inherent in the human condition. It is critical that we move toward seeing population growth as a product of the intricate interrelationships between birth rates, child survival, economic development, education, and the economic and social status of women. The United States, through partnerships with other nations and with international organizations, must support and participate in the scientific and technological research needed by international population programs to better understand both the environmental dimensions of the problem, and complex approaches to its resolution. ■

Electric Vehicles: Opportunity or Problem?

Based on remarks delivered to the New York Academy of Sciences on October 17, 1995.

ROBERT A. BELL

Vice President, ConEdison

N othing is ever simple. Innovations often present both opportunities for change and improvement and problems for existing relationships and societal priorities. Electric vehicles, and the various means for encouraging their use, are no exception.

Competitive Implications of Mandates

It is very important to understand the competitive implications of mandates for electric vehicles. The mandate originally envisioned by the Ozone Transport Commission would have required that 2% of all light-duty vehicles sold must be zero-emission vehicles in all states from Virginia to Massachusetts. That mandate would have increased the proportion to 5% in 2001 and 10% in 2003. If the vehicles are not in demand, however, and a mandate applies in one

> If zero-emissions mandates are put in place in New York and not in other states in the region, it will have negative effects on economic competitiveness. It will simply add another cost to doing business in New York.

state and not in a contiguous state, the mandate becomes an economic burden. It becomes a cost of doing business that is not shared across state boundaries. If, for example, New York implemented mandates and New Jersey did not, New Jersey businesses would not have to bear the financial burden of retooling transport vehicle fleets, and New York businesses would be sorely tempted to move to New Jersey. For electric utilities in New York, which sell electricity to businesses, this would be among the worst of possible outcomes.

It is critical to understand the long-term implications of mandate strategies before embarking on legislation.

Battery Research—United States Advanced Battery Consortium

One of the most important technology problems impeding the spread of electric vehicles is the state of research on batteries. The largest research effort is being led by the United States Advanced Battery Consortium, which is sponsored by the U.S. Department of Energy and comprises the auto industry, especially Ford, General Motors, and Chrysler; electric utilities; and battery suppliers. The research program is funded at about a quarter of a billion dollars and is divided into two parts.

The mid-term battery program is focused on gradually increasing the driving range of electric vehicles in all types of weather and temperature conditions. The long-term program is targeted at strategies to develop batteries that will allow electric vehicles to perform at a level near that of gasoline-powered vehicles. In either case, however, the research is not likely to produce marketable

results by 1998. We just do not yet
have the research answers that will
allow us to solve the battery problem. As the U.S. Government
Accounting Office noted in 1995,
"the benefits of long term electric
vehicle batteries are as yet uncertain, while the benefits of mid-term
batteries are unclear although their
feasibility has been demonstrated. .

> **If current vehicle purchas-
> ing patterns continue, the
> emissions benefits of elec-
> tric vehicles are going to
> be swamped by the trend
> toward consumer prefer-
> ence for higher emission
> light trucks and vans.**

. [H]owever, it remains unclear whether the feasibility of a long-
term battery will be demonstrated."[1] Rushing to mandates will
force the production of cars that do not have the capacity that the
market expects, and hence will both disappoint the market and
force the purchase and use of vehicles that will be less efficient
than gasoline alternatives.

Conflicts in Goals

It is important to recognize that, as much as everyone supports the
development of electric vehicles, there are and will continue to be
conflicts over goals in the short term (Table 1).

There are conflicts between interested parties. For example, utilities would like to sell the electricity needed to charge EV batteries.
All-electric vehicles would use the most electricity. Nevertheless, from
the point of view of the industry, hybrid vehicles, which use both
electricity and gasoline, may make more market sense. Interested parties are also experiencing internal conflicts. For example, within the
automotive industry, electric vehicles would compete with existing
product lines both in domestic and foreign markets.

[1] Government Accounting Office, *EV Update,* September 29, 1995, Washington, D.C.

Table 1. Conflicts over Short-Term Goals

	Goal	Conflict areas
Environmental groups	Clean air	Regional versus curbside impacts Costs versus effectiveness
Automotive industry	New markets	Competition with existing products Domestic versus foreign markets Hybrids versus all-electric vehicles
Utilities	New applications for electricity	Regional competitiveness of business Hybrid versus all-electric market effects

U.S. trends reveal an even more significant conflict between where the market is actually going and anyone's goals regarding electric vehicles. As Figures 1 and 2 illustrate, sales of light trucks, sport utility vehicles, vans, and recreation vehicles are soaring as sales of passenger cars have declined. Yet, the "light truck" category of vehicle is much less fuel-efficient than the passenger car. Since the early 1980s, fuel efficiency in the nation has declined to 24.8 miles per gallon. Less efficiency means more emissions and poorer air quality.

These market trends are so pervasive and significant that they will overwhelm the air quality effects of electric vehicle deployment. The effects of zero emission vehicles will be swamped by the opposite emission effects of current market preferences.

> **The greatest benefit to New York would be a national policy of tighter emissions for all gasoline-powered vehicles. After all, New York is downwind of a great many of the other 49 states.**

Regulatory Choices

As the nation considers the alternative regulatory approaches to reduce motor vehicle emissions, it is

While light truck sales soar . . . it takes more fuel to go places

Figure 1. Sales of passenger cars and light trucks.

Figure 2. Fuel efficiency as measured in miles a gallon.

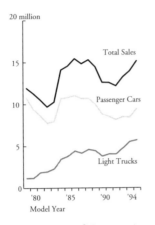

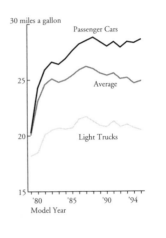

Source: Department of Transportation.

important to consider costs. What would it cost to reduce a ton of volatile compounds from the atmosphere using various regulatory approaches? The Washington research organization Resources for the Future has examined that question under two regulatory rubrics: command-and-control approaches, which use government mandates and requirements to change automotive behavior, and economic-incentive approaches, which provide incentives for change within the marketplace itself. The results are striking. Under command-and-control strategies, a nationwide mandate to deploy electric vehicles could cost as much as $108,000 per ton of volatile compounds reduced. Under economic-incentive strategies, using emissions levels to determine vehi-

> In a free-market economy, the key to hybrid electric vehicles is not regulation.

cle registration rates, with higher charges for registering higher emission vehicles, would cost under $2,000 per ton of reduced volatile compounds.

What is striking is not the absolute numbers, but the orders-of-magnitude differences. Reducing emissions and improving air quality is and must be a priority. But the approaches taken must carefully consider the price tags for alternative strategies, as well as their subsequent technological and market effects. ■

Electric Vehicles and the Automotive Industry: The Implications of Emission Mandates

Based on remarks delivered to the New York Academy of Sciences on October 17, 1995.

ROBERT STEMPEL

Executive Director, Energy Conversion Devices, Inc.

B attery-powered electric vehicles will become a key part of the nation's transportation system. Electric vehicles are quiet, fun to drive, pollution free, convenient, and require little or no maintenance. But the reason that electric vehicles are now emerging as a potential force in the automotive marketplace is not due to automotive innovation, but rather to technological innovation in the electronic and information systems industries.

From the early 1900s though the 1970s and 1980s, the auto

> **The reason that electric vehicles are now emerging in the automotive marketplace is not due to automotive innovation, but rather to technological innovation in the electronic and information systems industries.**

industry worked on electric vehicles. The results were, by and large, electric mechanical devices. Energy use was inefficient. Reliability was poor. Batteries were heavy, and energy density was low.

In 1987, however, innovation opened a new chapter and new possibilities. A solar-powered car, called the *Sunraycer,* crossed Australia at an average of 42 miles an hour for almost 2000 miles without ever fully discharging its batteries. The breakthrough was a combination of efficient design and computer-controlled power electronics. Computer systems allowed management of the energy collected from the sunlight to be used in the most efficient manner possible.

Interests of the Automotive Industry

Shortly after the *Sunraycer* achievement, General Motors began to develop a battery-powered vehicle to test the concept of computer energy management and solid-state power electronic devices. The result was the GM *Impact,* which operates on lead acid batteries and provides about a 60-mile driving range in the city and up to 80 miles on the highway.[1]

Why would a company like General Motors, whose entire business is premised on the internal combustion engine, be interested in electric vehicles? GM believes that their electric vehicles can occupy a key place in the household vehicle fleet as

[1]EDITOR'S NOTE: Shortly after this presentation, the *Impact,* now called *EV1,* was introduced commercially in Los Angeles, San Diego, Tucson, and Phoenix. The cars are selling well.

well as in a commercial fleet. For an urban commuter, electric vehicles will present an efficient, attractive option. For a commercial operator with a pre-determined delivery route, they will present similar advantages. Indeed, electric vehicles have proved their utility in the delivery business. United Parcel Service (UPS) operated electric delivery vehicles in Manhattan from 1950 until 1985. Each truck was really a warehouse on wheels, operating only 10 miles per day. But, given distances in Manhattan, this was perfectly acceptable. Nevertheless, even if all potential delivery fleet owners bought electric vehicles, the market would still be too small to be attractive. To achieve sufficient volume, the market must extend to the household.

> **The electric vehicle is more than just a car; it is a consumer electronic device. Potential customers have very high expectations of electronic devices.**

The second reason electric vehicle innovation is important relates to foreign competition. There is fierce competition between U.S. and Japanese automakers over who will be the first to market an electric car. The United States occupies a market leadership position in power electronics and computer technology. If that leadership is not applied to transport, the nation will miss a competitive opportunity. But, whether we are trying to reach American households or Japanese markets, improved technology will be key.

What Happens to Technology with Mandates?
Much of the electric vehicle experimentation to date has resulted from an industry-wide recognition of the potential of the market and a series of technological innovations. The technology is far from perfected; indeed, it is changing rapidly. What will happen to

> **Mandates will force production, but they will not assure sales. Mandates will actually set back electric vehicle development.**

innovation if large markets, such as California or New York, mandate that a percentage of manufacturers' products be zero-emission (which means electric) by 1998 in order to be able to sell gasoline vehicles in that market? I believe the answer will be a drop in innovation, a shrinking of the potential market, and, ultimately, a fall-off in interest in electric vehicles within the automotive industry. Why?

First, we simply do not have enough information to meet such a mandate at this point. We do not know enough to lock in on a single vehicle technology and have the industrial capacity for production in sufficient volume by as near-term a date as 1998. Current technology is going to become obsolete very quickly. For example, in power electronics, reliability is improving every day, costs are going down, and control algorithms are getting better.

Battery technology provides another example. A great deal of experimentation with battery alternatives is directed toward increasing speed and range. A recent nickel/metal hydride battery is now to the point that it can operate a car on the highway at between 45 and 60 miles per hour and achieve 200 miles on a single charge. In the city, it can regularly get 100 miles per charge.

But none of these innovations is ready for a mass market. The critical need now is to push those improvements out farther, not to lock a single improvement into a large production process. If we do not continue up the innovation curve, we will not have an ultimately successful vehicle.

Second, with a less-than-optimal product, demand will wither. The electric vehicle is more than just a car; it is a consumer electronic device. Today, consumers live and breathe electronic devices. They are used to taking them out of a box; switching them on; and having them operate immediately, quietly, and efficiently, with a minimum of time spent reading the owner's manual! Consumers are used to low or no maintenance for their electronic devices, and count on reliability and durability. Younger consumers may own two or three computers. They have no fear of electronic devices. Potential customers are going to have very high expectations of electric vehicles. If mandates result in underdeveloped products being rushed to market, those expectations will be disappointed. And every market analyst knows what happens to products that fail to meet consumer needs.

> The objective should be to develop infrastructure and to create several demonstration programs to launch the technology to achievable public expectations.

In short, mandates will force production, but they will not assure sales. If major manufacturers are forced to market to uncertain demand with predetermined sales volume, technology will be under-developed and customers will reject the product. Mandates will actually set back electric vehicle development.

Partnership Philosophy Needed

What should we do? A critical element of the electric vehicle strateg should be a partnership between the auto industry, the electric pow utilities, and the government. The objective should be to develop infrastructure and to create several demonstration programs to laur the technology to achievable public expectations. Government pol is critical here. Governments can provide incentives to stimulate e

purchase of vehicles by retail customers, to encourage the development of public charging networks by public utilities, and generally to expand and deepen the market. ■

Getting from Then To Now: A Brief History of the Electric Vehicle

Based on remarks delivered to the New York Academy of Sciences on October 17, 1995.

VICTOR WOUK

Principal, Victor Wouk Associates

E lectric vehicles were among the first automotive innovations in the United Sates. In fact, they even came complete with flower pots to decorate the door posts! Ironically, the early electric vehicles were put out of business by an electric device—the electric starter. The automotive owner no longer needed a crank and a set of well-developed upper body muscles to start the engine. Voila! The automotive age.

Automotive Innovation and Emissions

The automobile now represents freedom—freedom to go where you want to, when you want to, in a reasonable time frame, at a reasonable price, and with amazing reliability. Unfortunately, that freedom is accompanied by a complication: emissions that cause air pollution.

> **Renewed interest in electric vehicles has given rise to a new series of objections.**

Emissions have been reduced enormously over the last two decades. Indeed, federal mandates and new technologies have now brought emissions of a new vehicle to one-twentieth the level of 1968. But progress on a per unit basis does not always translate neatly into equivalent environmental impact. More and more vehicles are taking to the road, and they are being driven more and more miles. When more vehicles and more miles are combined with urban sprawl and, in a place like Los Angeles, geology, cleaner cars cannot tip the air quality balance.

Hence, there has been a renewed interest in electric vehicles. In their purest conception, such vehicles use no fossil fuels in their operation, and hence produce no emissions and no pollution. But, the renewed interest has also given rise to a new series of objections.

The Rise of Objections

On the environmental side, one of the most often-raised objections is that electric vehicles are really not pollution-free; they simply shift the source of the pollution from the tailpipe of the car to the smokestack of the utility that produces the electricity that powers the car. The arithmetic simply does not support that fear. Electric vehicles would use only a small fraction of the electricity generated by most utilities.

On the market side, the fundamental objection is that the vehicles will simply not go far enough fast enough. The early innovators thought if they built a car that would free the consumer from the cost of gasoline, that they would sell like hotcakes. Well, they sold like cold-cakes. The cars had an extremely limited range and merely crept along. Not to be discouraged, innovators tried to use efficient power electronics instead of electromechanical switches for speed controls, but that only increased the range from 35 to about 40 miles per battery charge.

So, the question was raised, Was there a better battery than the lead battery for resolving the speed and distance problems? In 1963 Dr. Lee A. DuBridge, President of the California Institute of Technology, became so fascinated with the problem that he organized a mini-seminar of chemists, chemical engineers, physicists, and electrical engineers to consider the alternatives. In short, they determined that several potential electrochemical reactions would have much more energy density than lead batteries, even though the result would still fall short of gasoline energy.

Dr. DuBridge's insights were well founded. Today, batteries available to power electric vehicles are three times as efficient and powerful as those twenty years ago. Future batteries now in the experimental stage will be five times as good. Technology has turned the corner on the battery problem.

> In the early days, the prediction was that electric vehicles would sell like hotcakes. Instead, they sold like cold-cakes. Speed and distance were serious problems.

> New technologies have now brought emisions of a new vehicle to 1/20th the level of 1968. But progress on a per unit basis does not always translate neatly into equivalent environmental impact.

Furthermore, indications are that consumers have too. Surveys of customers of the General Motors electric car, *EV1,* show that buyers value electric vehicles for their own merits, irrespective of environmental impact. They like not having to stop at the gas station. They like the reliability of just plugging the car in at night and going in the morning. They are satisfied with performance and reliability. Improvements in power and distance will only serve to reinforce these views. If the environment-based regulatory process can demonstrate sufficient patience,[1] technological innovation will produce electric vehicles that will appeal to a broad global market. ■

[1]Two months after the New York Academy of Sciences' policy meeting on the technological effects of mandated emissions regulations, the California zero emissions manates were deferred and subsequently postponed until 2003.

Conflicts Over World Population: Cairo and Beyond

Based on remarks
delivered at the New
York Academy of
Sciences on October 5,
1994. © 1995 by Joel
E. Cohen. Portions
of this text are based on
How Many People
Can the Earth
Support? by Joel E.
Cohen, published in
1995 by W.W. Norton.

JOEL E. COHEN
Professor of Populations, The Rockefeller University

P opulation and prosperity will remain important issues, and will require increasing attention, for decades to come. The United Nations International Conference on Population and Development (ICPD) held September 5–12, 1994, in Cairo, Egypt, aimed to build an international consensus about these issues, and succeeded in part. In this presentation, I shall sketch the context of the Cairo conference; describe some of its main achievements; and suggest critical conflicts that remain to be resolved.

The Context of Cairo

Compared to history before World War II, the context of Cairo—in fact, the entire human situation—is unprecedented in four respects:

- first, the size and speed of growth of the human population;

- second, the human impact on the physical, chemical and biological environment, and human vulnerability to changes in the environment;
- third, the enormous wealth of some parts of the world and the resulting disparities between the rich and the poor; and
- fourth, the cultural implosion that brings diverse traditions into contact, and sometimes into conflict.

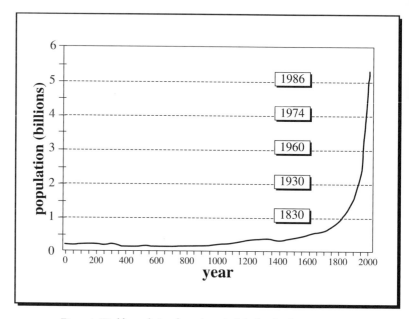

Figure 1. World population from the end of the last Ice Age

Population Growth: The Big Speed-up

Some 12,000 years ago, after the last Ice Age, the human population of the Earth first exceeded five million people. By A.D. 1650, the population grew to about 500 million, or half a billion. This 100-fold increase represented a doubling about once every 1,650 years, on the average.

Since A.D. 1650, population growth accelerated tremendously.

> **Never before the second half of the twentieth century had any person lived through a doubling of global population in a single lifetime—and now some have lived through a tripling of human numbers.**

The human population increased from roughly half a billion to roughly 5.5 billion today—three and a half doublings in three and a half centuries, or one doubling per century. Since World War II, the population has doubled in about forty years, a forty-fold acceleration over the average population growth rate prior to 1650. Never before the second half of the twentieth century had any person lived through a doubling of global population in a single lifetime—and now some have lived through a tripling of human numbers.

The populations of some domestic animals have grown even faster than human numbers. In 1990 to 1992, people had 4.3 billion large domestic animals, from sheep to camels. The number of chickens, 17 billion, more than doubled over the prior decade. In 1992, domestic animals were fed 37 percent of all grain consumed. Some of these domestic animals have major environmental impacts. They produce methane and liquid and solid wastes, overgraze fragile grasslands, and prevent forest regeneration.

The human species lacks any prior experience with such rapid growth and large numbers of its own or of its domestic species.

Environmental Impact: Rising Vulnerability

Humankind is now a large actor on the small stage of this planet. In the minds of many, people are besieged by an unprecedented litany of environmental problems, including loss of topsoil, desertification, deforestation, dropping water tables, toxic poisoning of drinking water, oceanic pollution, shrinking wetlands, overgrazing, loss of wilderness areas and species, shortage of firewood, siltation in rivers and estuaries, encroachment of human habitat on arable land, erosion of the ozone layer, global warming, nuclear wastes, and acid rain. Vulnerability to a rise in sea levels rises with the tide of urbanization, as the number of people who live in coastal cities rapidly approaches one billion.

Like the sudden giants in H.G. Wells' novel, *Food of the Gods,* humans have become a geological force. For example, in 1991, human use of inanimate energy was 93 billion megawatt-hours per year, up nearly 100-fold from inanimate energy use in 1860. The current level of inanimate energy use is nearly four times the total solar energy available to the Earth for human food production. Again, all human-made water reservoirs and dams today have a useful capacity of 3,000 to 5,000 cubic kilometers, roughly twice the stock of water in all the world's rivers.

Environmental vulnerability affects human health. With increasing frequency, people make contact with the viruses and other pathogens of previously remote forests, and new diseases are emerging.

In 1990, of the 4 billion people, or 77 percent of world population, who lived in developing countries:

- 1.5 billion people lacked access to health services

- 1.7 billion people (1 in 3 people on earth) were infected with tuberculosis

- 1.3 billion people lacked access to safe water

- 2.3 billion people had no toilet

- 1 billion adults (600 million of them women) were illiterate.

- Females received on average half the higher education of males

- Nearly 1 billion people chronically went hungry

Rich and Poor: Growing Economic Disparities

In the aggregate production of material wealth, the half-century since World War II has been a golden era of technological and economic wonders. For example, in constant prices, with the price in 1990 set equal to 100, the price of petroleum fell from 113 in 1975 to 76

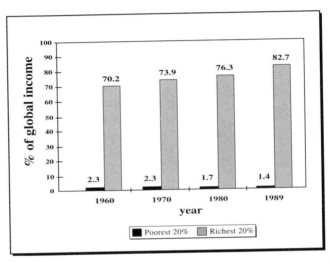

Figure 2. Shares of world income received by the 20 percent of people in the richest nations and 20 percent of people in the poorest nations, 1960-1989. Source: United Nations Development Programme, 1992.

in 1992. The price of a basket of 33 nonfuel commodities fell from 159 in 1975 to 86 in 1992. Total food commodity prices fell from 196 in 1975 to 85 in 1992.

But a rising average income has been very unequally distributed. In constant 1989 U.S. dollars, the absolute gap between the wealthiest fifth and the poorest fifth of the human population rose from $1,864 in 1960 to $15,149 in 1989.

In 1992, the 830 million people in the world's richest countries enjoyed an average annual income of $22,000—a truly astounding achievement. The almost 2.6 billion people in the middle-income countries received only $1,600. The more than 2 billion people in the poorest countries lived on an average annual income of $400, or a dollar a day. The 15 percent of the population in the world's richest countries enjoyed 79 percent of all the world's income.

Dollars are not the full measure of human well-being. In 1990–95, while Europeans enjoyed a life expectancy above seventy-five years, Africans still had a life expectancy of fifty-three years—below the world average twenty years earlier.

While food prices have dropped by half, the bottom billion are so poor that they cannot exercise effective demand in world commodity markets; they are economically invisible.

Cultures in Collision

The cultural implosion of recent decades, while difficult to quantify, is the change that is potentially most explosive. Migrations within countries and between countries, business travel, tourism, radio, television, telephone, fax, the Internet, recordings, printed media—all have shrunk the world stage. Terrorism and religious fundamentalism reach across former cultural boundaries to threaten people everywhere—from the poorest regions of poor countries to the towers of Wall Street.

ICPD

In spite of terrorists' death threats and last-second political skirmishing, some 15,000 people from 183 countries gathered in Cairo in September 1994 with the announced aim of building a consensus among the world's nations on population and development over the next twenty years. The Cairo meeting followed intergovernmental conferences on population held in Mexico City in 1984 and in Bucharest in 1974.

The political history of these three conferences roughly approximates the square-dance step known as "swing your partner." In this step, you and your partner, pulling against each other, exchange positions, then come back to your original positions.

To simplify greatly, the United States and some other wealthy countries went to Bucharest in 1974 to promote family planning, with the intent of lowering population growth rates and thereby making it easier for poor countries to develop economically. But the developing countries rallied to the slogan, "Development is the best contraceptive." They wanted the rich countries to transfer the capital assets and technology they required for economic development. Slower population growth would follow, they argued.

Ten years later, the positions were reversed. The U.S. delegation argued that if the developing countries would let free markets flower, their economies would develop and eventually their fertility would fall. However, with ten years' more experience of rapid population growth, the developing countries urgently desired support for family planning programs. The U.S. stopped its support of the U.N. Fund for Population Activities to protest what it charged were coerced abortions in China.

In the decade since 1984, nearly all countries have recognized very rapid population growth (faster than 2 percent a year) as a problem. But scholars still debate how best to slow population growth. The promoters of family planning programs argue that "Contraceptives are the best contraceptive." Others emphasize the economic, educational, health, and cultural factors (like the status of women and improved child survival) that make parents want to bear fewer children. Because there was no consensus among scholars about the most effective means of lowering fertility before the meeting, one could hardly have expected the ICPD to take a single clear direction. And it did not.

The Programme of Action produced by the Cairo conference was a mix of dream and sermon, of wish and prayer. There was something for everybody. By one count, the Programme contained more than a thousand recommendations. Of these, only a handful dealt with the desirability and means of reducing fertility and slowing population growth. The rest urged governments to improve almost every aspect of human well-being, but specified no priorities.

While abortion, adolescent sexuality, homosexuality, and extramarital sexual relations absorbed attention and generated a tremendous amount of rhetoric, the three main accomplishments of the ICPD are simple and important.

Family planning and contraception were placed firmly in the mainstream. Access to family planning was recognized as an individual right and a governmental obligation.

Women took center stage. The right of women to limit their own fertility was affirmed. Their health, jobs, credit, education, property rights, and reproductive autonomy emerged as primary concerns.

The U.S. resumed leadership in promoting slower population growth to enhance human well-being. President Clinton requested $585 million for population activities in fiscal 1995. Both houses of Congress approved $10 million more than the president asked for. The U.S. will become the largest single donor for population activities.

Issues for the Future

Very few analysts think that the next 350 years will see a further eleven-fold increase in human population like the last 350 years. The long-term demographic future of our species will not resemble its long-term past. An end to long-term average population growth is inevitable, very probably within the 21st century. The big questions are: just how soon, by what means, and at whose expense? Implicit in these simple questions are eight thorny issues that remain to be resolved:

1. Who will pay for family planning and other population activities? How will the bill be distributed between developing countries (who now pay 80 percent) and richer countries?

2. Who will spend the money, and how? How will monies be allocated between governments and non-governmental organizations? Between family planning and allied programs like reproductive health?

3. How will environmental goals be balanced against economic goals? If reducing poverty requires increasing production in developing countries, can the increased production be achieved at acceptable environmental costs?

4. How will cultural change be balanced against cultural continuity? In some cultures, the notion of empowering women contradicts directly the persistent call in the ICPD document for "full respect for the various religious and ethical values and cultural backgrounds." U.S. women achieved the vote only in 1920 and only after considerable struggle. Demanding equality for women asks some cultures to make far greater change in far less time. Such demands should be made with a clear and sympathetic understanding that they entail profound cultural change.

5. How will the often-asserted right of couples and individuals

to control their fertility be reconciled with national demographic goals if the way couples and individuals exercise that right happens not to bring about the demographic goals?

6. How will national sovereignty be reconciled with world or regional environmental and demographic goals? The control of migration, reproduction, and economic activities that involve the global commons (atmosphere, oceans, international water bodies, plant and animal populations) could easily generate conflict.

7. How will the desire and moral obligation to alleviate poverty and suffering in the short term be reconciled with the use of local scarcities as an efficient market signal? How can market economies meet the immediate needs of the world's poor?

8. On this finite sphere, how will rapid population growth and economic development in poor countries be balanced against high consumption per person in the rich countries?

These questions will not be answered easily, or perhaps at all. But they will persist for decades as a troubling heritage for our children. ■

The Economic Promise of Environmental Technologies

MARK SCHAEFER

Assistant Director for Environment, Office of Science and Technology Policy, Executive Office of the President of the United States

Based on remarks delivered at the New York Academy of Sciences on September 14, 1994. In 1995, Dr. Schaefer moved to the U.S. Department of the Interior, where he is a Deputy Assistant Secretary.

T he facts are easy enough to understand. We live in a world of more than 5.5 billion people with finite physical and biological resources. The global population is growing rapidly, placing even greater stress on already degraded resources. What is the likely outcome?

Pessimists can paint a harrowing picture: degraded waterways worldwide, unsanitary living conditions, steadily declining forested land, loss of biodiversity at accelerating rates, heavily polluted air in urban areas. It is a sorry image of failed industrialism, growth gone awry, and decrements in quality of life as a result of short-

> **Pessimists can paint a harrowing picture of failed industrialism, growth gone awry, and decrements in quality of life as a result of short-sighted economic gains. If the pessimists prevail, we will have passed along a world that none of us would want our children to live in.**

sighted economic gains. If the pessimists prevail, we will have passed along a world that none of us would want our children to live in.

The Preferable Scenario

In 1987, the World Commission on Environment and Development, more commonly known as the Bruntland Commission, articulated an alternative view—a forward-looking, optimistic scenario founded on two fundamental objectives: economic growth and environmental quality. Together these two objectives represent the cornerstones of a new global approach to the future: "sustainable development." In the words of the Commission, sustainable development is "development that meets the needs of the present without compromising the ability of future generations to meet their own needs." Implicit in this idea is the assumption that the global economy can grow steadily if resources are utilized in a sustainable fashion.

While the general concept of sustainable development is likely to be debated, redefined, and tested at its vague philosophical edges for years to come, its core principles should stand the test of time. The real question is whether 50 years from now, sustainable development will be viewed as the watchwords that awoke a profligate global community and set it squarely on a path of continued improvement in quality of life, or whether it will be eulogized as the much criticized and largely unheeded words of the wise. Will the promise of sustainability simply become a vaguely remembered goal in a commission report that went on to collect dust on the

world's bookshelf of lost causes?
Since the answer to this question
will define the world that our
children will live in, we must
hope not.

> **While the general concept of sustainable development is likely to be debated, redefined, and tested for years to come, its core principles should stand the test of time.**

Critical to the simultaneous
pursuit of both economic growth
and sustainable resource use are
environmental technologies—those technologies that advance sustainable development by reducing risk, enhancing cost effectiveness, improving process efficiency, and creating products and processes that are environmentally beneficial or benign.

The Bridge to a Sustainable Future

Environmental technologies fall into four major classes—*control:* the traditional "end-of-pipe" technologies that limit effluents into the water or emissions into the air; *remediation and restoration:* technologies used to redress environmental contamination and rejuvenate damaged ecosystems; *monitoring:* devices to track toxic substances in industrial systems, the distribution of pollutants in air, land, or water, or large-scale changes in the global environment; and *avoidance or prevention:* technologies that limit the production of pollutants or wasteful practices that degrade the environment or consume resources in an unsustainable fashion.

To expect industry to design and manufacture environmental technologies as an act of environmental and societal altruism is unrealistic. Although the public good is a consideration, economics is the ultimate driver. Hence governmental policies aimed at the public good must be devised in ways that foster innovation and commercialization by harnessing market forces. Where these forces fail to ensure the health and safety of the public and protection of

> It is unrealistic to expect industry to design and manufacture environmental technologies solely as an act of environmental and societal altruism. Governmental policies must foster innovation and commercialization by harnessing market forces.

the environment, regulatory actions are necessary. Moreover, when carefully crafted, many traditional command and control regulatory actions also spur technological innovation. A central challenge of future environmental and economic policymaking will be to determine the proper mix of incentive and control-based regulatory actions that will drive the development, application, and diffusion of environmentally benign or beneficial technologies.

Environmental technologies are the bridge to a sustainable future, to a future economy and resource use practices that will ensure a high quality of life for future generations. The path toward this future promises tremendous near-term as well as long-term economic benefits.

Presently, global markets for environmental technologies are conservatively estimated to be more than $300 billion. That estimate is based on a narrow definition that focuses on control, remediation, and some monitoring technologies. A broader definition that includes industrial processes and energy production technologies that prevent or minimize wasted materials and pollution would result in a considerably higher estimate.

World markets for environmental technology are projected to grow rapidly over the next decade, offering tremendous opportunities to those companies that can provide products, services, and low or no waste industrial processes at competitive prices. As shown in the table overleaf, the U.S. market is now, and is likely to continue to be, the largest source of global revenues for environmental technologies. However, growth in Latin America, Canada, and Eastern Europe and the former Soviet Union is projected to outpace growth in the United States over the period from 1992 to 1997. This pattern of growth offers both challenges and opportunities to U.S. industry.

In a recent report, *Promoting Growth and Job Creation Through Emerging Environmental Technologies,* the National Commission on

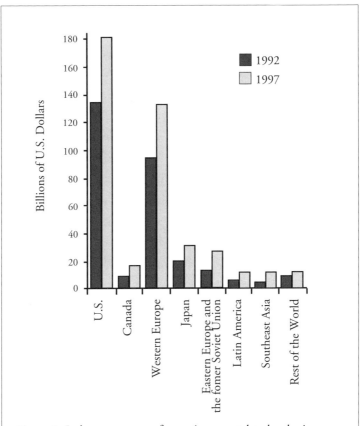

Figure 1. Industry revenues for environmental technologies. *Source:* National Science and Technology Council in *Technology for a Sustainable Future,* OSTP, 1994.

Employment Policy points out that more than one million workers were employed directly by the environmental technology industry in 1994. In addition, roughly an equivalent number of workers were employed in jobs that depend indirectly on this industry. According to the Commission, the annual rate of direct job creation in this industry is "more than double the average annual rate of growth in total employment for the U.S. economy in recent decades." If U.S. companies that develop and deploy environmental technologies can maintain a strong competitive position in the years ahead, the nation will reap tremendous benefits, including the continued growth of high-quality jobs for American workers.

Yet the benefits of these technologies go well beyond readily quantifiable near-term economic considerations. The long-term reward will be a higher standard of living and sustained economic growth in a world of finite resources: sustainable development.

Population Growth and Technological Innovation

At the end of World War I the world's population was 2.5 billion—today it is more than 5.5 billion. On the basis of medium growth rate assumptions, the United Nations projects a world population of 8.5 billion by the year 2025. Most of this growth will occur in developing countries. World population growth rates of this magnitude could have major adverse effects on global natural resources unless consumption and conservation practices change significantly. Improving, or at least maintaining, the present standard of living in nations throughout the world will require large-scale application of environmentally sound technologies—technologies that will enable the sustainable use of resources.

The population summit in Cairo in 1994 and the earth summit in Rio de Janeiro in 1993 highlighted the connection between rapidly increasing population growth and potential environmental degradation. According to U.S. Undersecretary of State Tim Wirth: "Everything we can do for political stability may be overwhelmed by sheer numbers of people. What can we do economically without stabilized populations? If we are concerned about the state of the Earth which sustains our species—if we are to

> The availability of clean technologies is by no means the sole answer to national or global environmental problems. Human behavior—the collective choices made on a daily basis by billions of individuals—will ultimately determine whether sustainability is achieved.

have any chance of sustaining it—population stabilization is a vital goal."

Urban areas, particularly those in developing countries, are especially vulnerable to unconstrained growth. In 1990, 40 percent of the world's population lived in urban areas. This proportion is expected to grow to 60 percent by the year 2025 according to UN projections. In 1990, 13 cities worldwide had populations of 10 million or more. By 2010, this number is likely to double. Nearly all of these megacities will be in developing countries (see Figure 2).

Major urban centers face particular challenges in ensuring adequate supplies of clean water, properly treated sewage, a dependable energy supply, and a well constructed infrastructure. In the decades ahead, the appropriate application of environmental technologies will be key to achieving a desirable quality of life in major urban centers worldwide.

The availability of clean technologies, however, is by no means the sole answer to a nation's environmental problems. Human behavior—the collective choices made on a daily basis by millions of individuals—will ultimately determine whether sustainability is achieved. Relatively minor changes in consumption practices by a significant proportion of society can lead to dramatic changes in

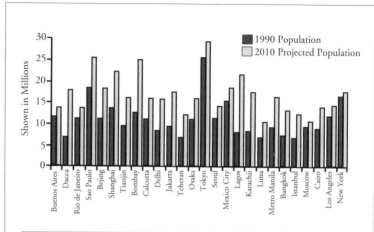

Figure 2. In 1990, 13 urban areas in the world had populations of 10 million or more. In 20 years, it is estimated that there may be 26 of these megacities, according to a UN medium-growth rate scenario. These huge urban agglomerations face significant environmental problems. *Source:* National Science and Technology Council in *Technology for a Sustainable Future*, OSTP, 1994.

resource and technology use. Increasing consumer preferences for products made from recycled materials or for more energy-efficient homes, for example, has spurred entirely new industries, manufacturing processes, and products. In the future, industries that provide products and services that minimize the use of virgin materials and conserve energy are likely to have a decided market advantage as resources become increasingly scarce and, hence, more expensive.

The Path to Sustainability

Technology is seen by many as the source of—not the solution to—national and global environmental problems. Indeed, there are many examples of short-sighted technological developments that have led to environmental damage: from industrial pollution associated with poorly conceived manufacturing processes, to inefficient energy production practices with inadequate pollution controls. Yet manufacturing and energy-use technologies, systems, and practices have changed dramatically over the past quarter-century. Automobile technologies are a case in point. Since 1970, carbon monoxide and hydrocarbon emissions per vehicle have decreased dramatically. This is the good news. The bad news is that a four-fold increase in vehicle miles traveled since 1960 has largely offset these reductions, making it difficult for many U.S. cities to meet air quality standards. This example illustrates the power and limitations of technology, as well as the importance of considering population growth and consumer practices when devising public policy.

Twenty-five years ago Barry Commoner wrote in his landmark book, *The Closing Circle:* "The chief reason for the environmental crisis that has engulfed the United States in recent years is the sweeping transformation of productive technology since World War II. The economy has grown enough to give the United States population about the same amount of basic goods per capita as it did in 1946. However, productive technologies with intense impacts on the environment have displaced less destructive ones. The environmental crisis is the inevitable result of this counter-ecological pattern of growth." Today we can point to signs of a reversal of this trend. Increasingly, industry is developing, and consumers are purchasing, technologies and other products that are more energy efficient and generate less waste. Through ingenuity, commitment, and enlightened public policies, we may, in the near future, be able to point to a global technological transformation that pulled the world away from the brink of environmental crisis.

The Clinton/Gore Administration has worked for the past several years to ensure that federal policies encourage the development, commercialization, application, and export of environmental technologies. In July 1994, the Administration published

Technology for a Sustainable Future: A Framework for Action, which describes the key policy issues associated with the advancement of these technologies. This report served as the foundation for more than two dozen workshops and conferences held throughout the country involving hundreds of individuals from the private sector and the state governments. These discussions led to a second document, *Bridge to a Sustainable Future,* which lays out the key elements of a national environmental technology strategy. As President Clinton and Vice President Gore jointly state in the preface of the strategy report: "Now the time has come for creative action and bold steps. Let us pledge to use technologies wisely for they are the bridge to a sustainable future. Our foresight will define the structure of that bridge. Our creativity will allow us to build it. And our commitment will determine how quickly we cross it."

There is a critical need to continue the process of technological transformation of U.S. industry by putting policies in place that foster innovation and competition. At the same time, we must encourage and aid developing nations in making a similar transformation to help them achieve long-term economic growth, resource protection, and continued improvement in quality of life. If we are successful, new ideas, new products, and new services will be deployed both in the domestic marketplace and throughout the world—and we will be squarely on the path to a sustainable future. ■

III

Spurring Growth

Whether you surf the net in Tokyo, punch a time clock in Detroit, craft software in Bangalore, or manage billions of dollars of assets in New York, the reality of the vital link joining the mastery of knowledge, technological innovation, and economic growth is everywhere ascendant. Yet there is considerable distance between knowing that something is so and knowing how to make it happen. Sorting out how to best link technology investments to economic growth targets, and ensuring that benefits are maximized, can be difficult. Indeed, deciding what emphasis to give technology policy at all relative to the macroeconomic setting remains a major puzzle. However important technology may be on the shop floor and in the laboratory, its relationship to economic policy priorities can be less clear. The essays in this section address these difficulties in terms of private industry, government initiatives, and military–civilian economic relationships.

Innovation at Risk: The Future of America's Research-Intensive Industries

Based on remarks delivered at the New York Academy of Sciences, November 7, 1995.

J. Ian Morrison, *President,* and William G. Pietersen, *Chairman*
Institute for the Future

T he new reality of the global marketplace is that you are not competitive unless you are globally competitive. The correlation between innovation, competitiveness, and growth is direct and relentless. America has enjoyed high rates of innovation, and hence competitiveness, because it has traditionally been among the world's leaders in R&D spending. That is changing, and the implications for the future are not pleasant.

The Importance of the Eight

The future growth of the American economy and the prosperity and well-being of the American people are closely tied to the suc-

> **America's eight most research-intensive industries employ 5.2 million people, support an annual payroll of a quarter of a trillion dollars, and have final sales of nearly a trillion dollars.**

cess of the eight most research-intensive industries in the United States. These eight industries are chemicals, aerospace, communications, equipment, scientific instruments, semi-conductors and electronic equipment, software, pharmaceuticals, and computer and office equipment. These industries are "research intensive" in that their research spending exceeds the national industrial average.

The sheer scale of these industries within the American economy is striking. They employ 5.2 million people, pay nearly a quarter of a trillion dollars in payroll, and have final sales of nearly a trillion dollars. More striking still is their role in U.S. economic growth. Between 1953 and 1968, the annual growth rate in expenditures on research and development of these industries was about 8%, three times the rate of growth of the GDP. It was these investments that set the roots of current economic growth. In the last 25 years, the research-intensive industries have grown twice as fast as the economy as a whole, and they produce a net trade surplus of $50 billion. Together, these eight industries pay for 70% of the industrial R&D in the country, yet their combined sales account for only a third of net U.S. manufacturing sales. The intensity of that investment in research will yield unknown benefits in the future.

A Future in Jeopardy

Although these eight industries have been key to the U.S. economy in past decades, there is growing cause for concern about their future. As Figure 1 illustrates, the rate of growth in the nation's R&D commitments has slowed relative to the rate of growth of the GDP. Yet simultaneously, innovation is becoming ever more

important to economic advance.
There are eight driving forces that
underpin the nation's move away
from R&D commitments and the
consequent threat to the future.

> **Spending for R&D in electronic components has decreased to 25% of its level five years ago.**

First, the end of the Cold War
has removed the historically compelling rationale for making
national investments in research. In the last five years, annual
funding for industrial R&D for aircraft and missiles fell by half,

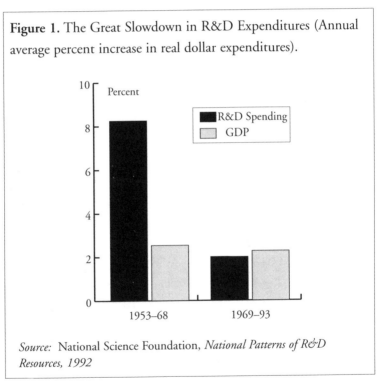

Figure 1. The Great Slowdown in R&D Expenditures (Annual
average percent increase in real dollar expenditures).

Source: National Science Foundation, *National Patterns of R&D Resources, 1992*

and spending for electronic components decreased to 25% of its
previous level. Although there is much criticism of current levels of
government support for defense R&D in light of changing geopolitics, the fact is that defense spending in the past has led to spinoffs

> **Costs of research are rising to a point where it is difficult for a single company to absorb them.**

in civilian innovation. This dynamic may not thrive in the future.

Second, global competition is squeezing R&D returns. Many large firms, especially those involved in high-technology global markets, are experiencing a decline in margins. The competition is not coming only from industrialized economies, but is truly worldwide in its origins.

Third, the costs of research are climbing. The costs of a new generation of aircraft are between $50 and $100 billion. Every new drug requires an investment in excess of $300 million. A new car model costs between $3 and $6 billion. A new generation of computer chips can cost over $1 billion in basic investment. The costs are rising to a point where it is difficult for a single company to absorb them.

Fourth, R&D has become more oriented toward product development. Industry is cutting back on basic research. The long-term consequences of this trend are unclear.

Fifth, R&D is becoming globally disseminated more quickly. In turn, this speeds the pace of competition.

Sixth, U.S. rates of personal and corporate savings are the lowest of any nation. This constrains the ability to invest in long-term research and development, especially given the extraordinary costs of the R&D process.

Seventh, institutional investors are incredibly focused on short-term performance. Wall Street's reaction to news of increased corporate investment in research has generally not been positive. Long-term commitments raise fears about short-term returns. When investors become nervous, corporations are more hesitant to reinforce their long-term plans.

Eighth, expanded regulation risks penalizing innovation. Price, market, and environmental regulations are all double-edged swords with the potential to cut dangerously into research-intensive industries.

> **The government should provide the support necessary to increase total national R&D investment from the current 2.6% of GDP to 3%, at least matching our main international competitors.**

An Agenda for Action

If the future is in jeopardy, what ought to be the response? It is critical that the nation increase its public and private investment in research and development to ensure continued growth and global competitiveness. The government should provide the support and context necessary to increase total national R&D investment from the current 2.6% of GDP to 3%, at least matching Japan and Germany, our main international competitors. Moreover, this may not be enough. Such a commitment will not put America in a leadership role; it will merely make the nation's commitment comparable to its counterparts.

Under this general objective, there are nine specific actions that need to be taken. First, government spending on basic research must be maintained. Support for basic research should be allocated 50% to defense and 50% to civilian needs.

Second, R&D tax incentives must be substantially and permanently increased to encourage high levels of private investment and longer term R&D.

Third, the tax rate on long-term capital gains should be lowered significantly, perhaps to 15%.

Fourth, government must rely on market forces to create a regulatory environment that sets clear compliance standards and that reduces government-generated risk associated with innovation.

> **The complexity and costs of research are rising, and innovative joint ventures and corporate collaboration will be necessary if the nation's industries are to stay competitive.**

This is particularly true for the regulatory approach to fledgling industries such as biotechnology.

Fifth, intellectual property rights should be expanded, and international enforcement aggressively pursued.

Sixth, tort system reform is needed to establish national standards for product liability and reduce the risk of litigation for innovative firms that invest capital in R&D ventures. This is particularly acute in some biomedical areas where liabilities can be enormous.

Seventh, antitrust policies and laws that discourage joint corporate ventures should be re-examined in light of the emerging global markets. The complexity and costs of research are rising, and innovative joint ventures and corporate collaboration are going to be necessary if the nation's industries are to stay competitive. A related eighth point is that a suitable legal, administrative, and infrastructural framework is needed to encourage cooperative research.

Finally, the government must continue to fund and encourage science education, especially at universities. Intellectual capital produced by U.S. universities is being deployed internationally. U.S. policy must continue to encourage and support science education in its own university system as the fundamental resource for America's own innovative enterprise. ■

Achieving Economic Growth

Based on remarks delivered to the New York Academy of Sciences on March 5, 1996.

WILLIAM J. MCDONOUGH
President
Federal Reserve Bank of New York

O ne of the most important underpinnings of a successful link between technology and economic development is developing and maintaining a long-term, noninflationary economic growth environment. A robust economic environment is essential both for the nation and for the New York region. There are four factors that are critical to that larger goal: a preemptive approach to monetary policy; price stability; deficit reduction; and promoting the economic and financial well-being of weaker segments of society.

Preemptive Monetary Policy

A "preemptive" approach to monetary policy means dealing with inflation before it takes root. One of the most heartening features of the recent period of economic expansion is the continuing good news about inflation. Monetary policy in early 1994 was the tool

> A robust economic environment is essential both for the nation and for the New York region.

that firmed up monetary conditions and hence reduced inflationary pressures before they could affect general prices. Why has monetary policy played such a key role in the economy?

First, it is important to appreciate that monetary policy works with uncertain and long time lags. Most of the effect of monetary policy on economic activity takes place over one to two years. But, its effect on prices can be up to three years or longer. Hence, effective monetary policy must anticipate the economy three years in the future in order to have an appropriate effect on inflation.

Second, the inflation experiences of the last two decades indicate that an overheating economy has a much stronger effect on fueling inflation than sub-par growth has on lowering inflation. Once an economy overheats, it is extremely difficult to rein in inflation. This asymmetry—a favorite central banker term—reinforces the need for preemptive monetary policy. Failure to contain inflationary pressures at early stages makes it much tougher to deal with inflation later. If you let inflation grow, the problem gets bigger and bigger and the costs to the American economy to bring prices under control get higher. We are very aware that there are real, live people in that economy who suffer if we fail to act on a timely basis to beat inflation down. Fortunately, the degree to which elected officials, economists, and the general public understand this asymmetry has grown significantly in recent years. So there is broader societal and political support for preemptive monetary policy, which in turn has contributed to its more effective use as a policy instrument.

Price Stability

The primary long-term goal of monetary policy for sustaining economic growth should be price stability. Price stability is achieved when inflation is not considered in household and business decisions. When inflation is under control, businesses and families have confidence in making spending and investment decisions. In turn, those decisions provide the fuel for economic growth.

> **The economic growth–price stability relationship provides a long-term anchor for monetary policy.**

Hence, I do not believe that there is a conflict between price stability and economic growth. Indeed, sustained economic growth is the means by which you achieve price stability. And only with price stability can the economy expect to achieve the highest possible levels of productivity, real income, employment, and living standards. The economic growth–price stability relationship provides a long-term anchor for monetary policy. Indeed, it is an appropriate measure by which society (and Congress) should hold the Federal Reserve Bank accountable for good or bad results.

A stable economic and financial environment certainly will enhance the capacity of monetary policy to fight occasions of cyclical weakness in the economy. That does not imply that we need (or have) targets for real economic growth. The Fed does not seek to limit growth to two or two and a half percent per year. We have no such targets. In trying to determine the extent of future inflation, we use a variety of economic indicators that reflect demand pressures and supply developments in the economy. These include labor markets, industrial capacity, utilization rates, estimates of the gap between actual and potential gross domestic product, developments in commodity prices and monetary aggregates, the extent of foreign competition, and the like.

> **When inflation is under control, businesses and families have confidence in making spending and investment decisions that provide the fuel for economic growth.**

Unfortunately, there is no straightforward summary measure that provides a reliable overall assessment of the many complex and diverse influences on inflation. Clearly, the complexity is such that one cannot have absolute growth "targets" with a simple link to monetary decisions.

The Federal Deficit

Even with price stability, economic growth, and monetary policy solidly linked, however, an economic blue sky is not ensured. The U.S. economy faces a very serious structural problem. In sum, we consume far too much and save far too little to sustain a healthy economy over the long term. At the household level, which I believe is the best perspective, we consume over 95 percent of what we earn. We save about 4.5 percent. That rate of savings is simply not enough to fuel the economic investment we need. So, what have we as a nation done? We have imported the savings of other nations to finance our own needs.

To make a bad matter worse, however, the federal budget deficit consumes much of the nation's total household savings.

That, by itself, is not necessarily bad. If we were China with a savings rate close to 30 percent, or Japan at 18 percent, or Chile at 20 percent, then we could afford to finance the deficit and have savings left over for economic growth. But using a large portion of our savings to finance the deficit simply leaves too little left over for productive investment.

We really do not have a formula for raising the American savings rate. How can we induce Americans to save more? Many strategies have been deployed. Few have worked. America's savings behavior appears to be a cultural problem that is poorly understood. That leaves reducing the deficit as the only alternative. It is critical to do not because it is fashionable, but because it is essential for the longer term health of the economy.

Societal Well-Being

It is important to recognize that taking budgetary action will have societal consequences. And policy makers must address these consequences. We may have less public money for many societal needs, but we must not walk

> **Deep inequities in the social fabric will lead to a poor prognosis for the national economy. No society in history has prospered with a continuously growing underclass.**

away from those needs. We must respond to them. Deep inequities in the social fabric will lead to a poor prognosis for the national economy and for the nation as a whole. No society in the history of the world has prospered with a continuously growing underclass.

It is clear that government alone cannot solve many of the problems we face; lasting solutions require community commitment. Private institutions must become deeply involved in solving local problems.

How does an institution like the Federal Reserve Bank of New York fit into such an equation? The New York Fed actually plays three roles in the Federal Reserve system. We have the major operating responsibility within the U.S. Fed system and help to design and implement monetary policy. Second, we are the international face of the overall Federal Reserve System. Third, we are responsible for the Second Federal Reserve District—this region. Although the national and international roles can be seen as all-consuming,

> **It is unacceptable that the New York region is growing its economy at a rate significantly below that of the rest of the nation.**

we have made a renewed commitment to our role in the region. We believe that it is unacceptable that the New York region—our region—is growing its economy at a rate significantly below that of the rest of the nation.

As an institution, therefore, we are increasing our capacity to analyze economic trends in the region and deepening our participation as a partner with community leaders who are designing programs to strengthen the economic development of the region. The November 1995 conference in partnership with the New York Academy of Sciences[1] was an expression of this institutional commitment. The New York Fed is also working with educators to develop courses on banking and is reaching out to high school and college students to improve their understanding of the role of banks in the economy.

In addition, Fed staff members, as individuals, are deeply involved as volunteers in the region as mentors, providing assistance to both students and adults on various banking and economics issues. The staff has worked with the community leaders to determine the needs, and then has stepped in to help. ■

---

[1] *In November 1995, the Federal Reserve Bank of New York and the New York Academy of Sciences co-sponsored a conference on Technology and Economic Development in the Tri-State Region, in collaboration with the Regional Plan Association, the Port Authority of New York and New Jersey, the Partnership of the City of New York, and the U.S. Department of Labor's Bureau of Labor Statistics for the New York Region. The proceedings of the conference were published as volume 787 in the Annals of the New York Academy of Sciences under the title* The Technology Link to Economic Development: Past Lessons and Future Imperatives.

Partnerships Linking Technology to Economic Growth: Case Experience from around the Globe

Summary of paper
delivered to the World
Congress of
Engineering Educators
and Industry Leaders,
Paris, July 3, 1996.
Research funded by the
Carnegie Corporation
of New York and the
John D. and Catherine
T. MacArthur
Foundation.

SUSAN U. RAYMOND AND RODNEY W. NICHOLS

Experimentation with ways to link science and technology policy to plans for economic growth has been widespread in the last decade. Policy-makers and private executives in economic regions, individual nations, and even urban areas, large and small, have come to recognize that, as the World Bank has noted, technological progress fuels productivity and "Growing productivity is the engine of development."[1]

A key philosophy running through all successful marriages of technology and economic development is partnership among industry, universities, and public policy. It is now widely recognized that

> **Vibrant partnerships cannot be freeze dried and mass marketed.**

it is both the "pull" of the market and the "push" of science and technology, all within a nurturing public policy environment, that have characterized the closest and most successful collaborative arrangements. What does it take to create such partnerships? What are the preconditions and what elements serve to increase the probabilities of success? An examination of the experience of several of the states of the United States, nearly all of which have now undertaken targeted technology policy initiatives, and that of many other countries from both the industrialized and industrializing world provides a starting point for understanding key ingredients for success.

Preconditions

Vibrant partnerships cannot be freeze dried and mass marketed. They are very much a product of the specific economic, cultural, and technological conditions of any particular geographic entity.

Social and Economic Context

It is important to recognize from the outset that policies targeted at creating partnerships rest within a larger context of social values and economic capacities. For example, the values embedded in different societies and cultures will affect not only choices among policies but also the perception of the importance of technology itself. Indeed, in some cases, technology-based initiatives may appear to be (and often, in fact, are) a threat to established economic or political power as well as a path to prosperity.

Overall economic policy, political stability, the rule of law, market competition, and myriad other macroeconomic policy decisions also set clear parameters within which science and technology policy can (or cannot) link effectively to economic objectives. Where tax policy is confiscatory, where property law is nonexistent, where monopoly is allowed to thrive, policies aimed toward harnessing corporate and university innovation to economic growth objectives will have little effect. Equally, feeble support for science and innovation can thwart economic initiatives. As Paul Romer has noted, "No amount of savings and investment, no policy of macroeconomic fine-tuning, no sort of tax and spending incentives can generate sustained economic growth unless it is accompanied by the countless large and small discoveries that are required to create more value from a fixed set of natural resources."[2]

The Necessity of a Long-Term View

Successful partnerships are neither born nor matured in a matter of months. Indeed, strategies require years—and often decades—to bear fruit. Japan's experience provides a classic example. Japan's knowledge-intensive economic strategy dates to the late 19th century. Even by the 1960s, Japan had combined dedicated public investment with a clear policy understanding that full returns on those investments would not be seen until the 21st century. In Singapore, a decade of investment in the National University of Singapore was needed to create high-technology research and development capacity. In Hong Kong, a decade and a half of commitment was required to bring the new University of Science and Technology to full operation. The patience to measure time in units of ten is an important context for technology–economic development partnerships.

Therein lies a difficulty which, ironically, is enhanced by the very global changes that call for science and technology links to economic growth in the first place. The rise of democratic systems around the globe has created electorates empowered to

> **The patience to mea-
> sure time in units of ten
> is an important context
> for technology–econom-
> ic development
> partnerships.**

hold public policy accountable for results. But electorates often do not take a long-term view. Action and results are expected to be immediate, and accountability is at the ballot box. In such an atmosphere, policy-makers may not be predisposed to patience. The obstacles that prevent firms, states, or nations from taking a long-term view extend beyond the electorate, however. In many developing countries, economic reform itself has shortened investment time horizons. Speculative and rent-seeking activities, focused on maximum gain in a minimum time frame, can become significant economic engines, crowding out technological investments that entail longer gestation periods for economic payoff. Even within thriving private R&D-based corporations, the pressure of global competition and the need to respond to markets interlinked by telecommunications as never before have compressed the time in which investments are required to pay off and provided disincentives for the long-term view.

Characteristics of Successful Partnerships

In many countries, the conduct of science and technology is a multisectoral matter involving governments, corporations, universities, and, in some cases, not-for-profit research institutes. Figure 1 provides an illustration from the United States of the relative balance of these various "stakeholders" by various measures of R&D expenditure.

Beginning from the Beginning

Most successful partnerships are premised from the very outset on

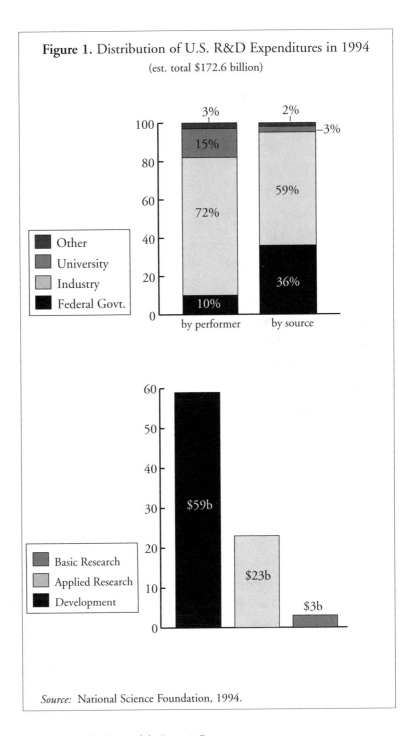

Figure 1. Distribution of U.S. R&D Expenditures in 1994
(est. total $172.6 billion)

Source: National Science Foundation, 1994.

> **Most successful partnerships are premised from the very outset on collaboration among institutions that all perceive mutual gain.**

collaboration among institutions that all perceive mutual gain. Business, universities, and government join together not simply to solve a problem, but to identify and diagnose the problem. Both the analysis and the agenda for action are jointly developed, jointly accepted, and jointly implemented.

In the United States, several recent cases illustrate this approach. In the state of Kansas, for example, Kansas, Inc. was created as a private, not-for-profit organization with a joint board of directors from the public and private sector to analyze the state's economy and develop plans for action, including technological programs. Although the impetus to create Kansas, Inc. was government led, the actual assessment of the Kansas economy and its technological dimensions was financed and led by private corporations and banks. The ideas generated by the analysis are then passed on to the Kansas Technology Enterprise Corporation (KTEC), also a private nonprofit organization with a joint public, private, and academic board of directors, to develop program initiatives.

In England, the experience of the city of Birmingham teaches a similar lesson about the importance of a partnership spirit from the very beginning of policy formulation. Between the late 1970s and the early 1980s, Birmingham lost over 200,000 jobs, 30 percent of its employment total. Employment in manufacturing fell by more than half. Gross domestic product per capita in the region fell from third highest in Great Britain in 1977 to second lowest in 1983. In response, the city council, the University of Aston, and the Birmingham Chamber of Commerce and Industry joined together to encourage the development of technology-based industries, including small businesses, and to foster links between the university and private business. The city acquired the land for a science park; the university upgraded its science and technology capability; private enterprise identified likely industrial investors and joined the city as a full partner in ownership of the National Exhibition Center designed to attract

> **Finding a balance between the culture and priorities of academia and those of industry poses a challenge in many settings.**

corporate trade groups to the city and introduce them to its technological capacities.

New York City has also joined the partnership strategy. Faced with a difficult recovery from the most recent recession and a serious deterioration of real estate occupancy in the Wall Street area, the city joined with local business and universities to analyze the needs of the area. An unsung and growing sector of New York's economy was the information technology and multimedia industry. Small, creative entrepreneurs needed high-tech but affordable space to plant roots and grow. But they also needed links to the city's world-class R&D capability in order to attract the brightest of young technological talent and to bring new research to market. Through partnership of both analysis and action, a new nonprofit corporation was formed, the Information Technology District Corporation, Inc. To the partnership, the city brought tax incentives, the private sector brought investment, and universities brought research access and expertise. The result is a strategy to create a "silicon alley" in New York, which includes a "vertical incubator," the InfoTech Center, which is the "hottest wired" technology center in New York.

It must be recognized that partnership strategies, which are premised on mutual self-interest and joint effort from the very inception, face difficulties in many economic settings. Where economies are dominated by small businesses without a significant R&D base, for example, finding private sector "partners" can prove problematic. In many settings, moreover, the public policy tradition is not one of shared decision-making and flexible collaboration with the private sector, but one of "command-and-control" regulation and economic planning. Seeing business and universities as valuable stakeholders in both problem analysis and strategic action runs against the current of much public policy tradition.

The Entrepreneurial University

A second characteristic of nearly every successful partnership is the presence of one or more "entrepreneurial" universities. The university partner sees itself not as an impenetrable academic fortress but as a dynamic player both on the S&T/economic policy stage and in the commercialization process.

In the United States, such roles are both historic and increasingly common. After the U.S. Civil War, the economy of the state of Georgia was totally devastated. Virtually every industry had been destroyed, little capital remained, and the markets for its traditional agricultural goods had disappeared. Business leaders and the state government, however, joined together to develop an economic recovery plan premised on technology. Georgia Tech University was conceived and born to be an entrepreneurial university, linked to industrial development through the quality of its research capability and the nature of its partnerships with private corporations for product development. More recently, the role of Stanford and Cal Tech in Silicon Valley, MIT in the development of Route 128, and the University of Texas in the creation of the technology initiatives of Austin, Texas all point to the growing importance of the aggressive university as a critical partner in successful partnerships.

The result in the U.S. has been a startling increase in the number of university–industry partnerships, stimulated and funded in part by the federal government. Figure 2 provides data on the growth of formal relationships over the last several years.

Entrepreneurial universities are key partners around the world. The National Tsing Hua University in Taiwan, for example, is an aggressive collaborator with technology companies located in its home city of Hsinchu. University faculty act both as consultants to local corporations and as sources of research innovations, which are then commercialized by those corporations. In France, one of the critical magnets for attracting corporate investment in Sophia Antipolis, the booming science park in southern France, was the commitment of the University of Nice to invest heavily in mathematics, computer sciences, and physics. The willingness of the university to become an aggressive high-tech research partner for corporations, as well as a mechanism for expanding the skilled labor pool, provided assurance to corporate managers that

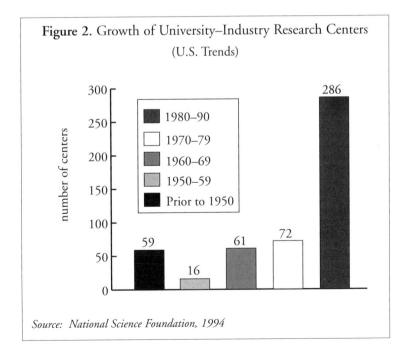

Figure 2. Growth of University–Industry Research Centers (U.S. Trends)

Source: *National Science Foundation, 1994*

a partnership with the region would be profitable.

In Japan, universities traditionally did not enter the commercial fray. Indeed, regulations prohibited corporate support for directed university research. Even today, faculty members generally are government employees, which results in conflicts over intellectual property rights. With changes in the corporate–university regulations, however, traditional academic detachment is shifting to closer industrial collaboration. Joint research projects between industry and universities numbered fewer than 100 in 1983; ten years later there were more than 1200 such projects.

Finding a balance between the culture and priorities of academia and those of industry poses a challenge in many settings. Concerns over publication rights, the setting of research priorities, access to intellectual property, and the continued health of the scholarly process have dampened the enthusiasm of many academic leaders about a more proactive partnership role. Yet, case experience indicates that the integrity of the academic enterprise and the pragmatic market-oriented priorities of industry need not be mutually exclusive. Partnership can lead not only to more competitive industry, but also to a renewed and revitalized university commitment to science and engineering.

Institutional Innovation

In most successful partnerships, new institutional vehicles are required to embody and implement partnership initiatives. No single sector—business, government, or academia—encompasses all of the networks and skills needed to link technology to economic growth. A collaborative organization is necessary, and it must embody a competitive attitude that will successfully respond to the marketplace along with the accountability that is incumbent upon any effort that involves government participation. Crafting innovative institutions that are credible to all partners and address each partner's need and motivations is essential to partnership success.

In the United States, a standard approach has been to create independent, not-for-profit umbrella organizations that integrate the interests and priorities of governments, universities, and private corporations, yet remain independent of any particular organization. These institutions normally are governed by a board of directors whose members are drawn from and are representative of industry, universities, and state government agencies. They act in their private capacity, not as representatives of their individual organizations, but reflect the views and interests of the sectors in which they work. The financing of the partnership organization can derive from several different streams of revenue, but usually is *not* linked to the state's general tax revenues. KTEC in Kansas, for example, is funded in part from a percentage of the revenues of the state lottery and in part from earnings for services it provides to business. The Montana Technology Alliance is a not-for-profit partnership technology foundation that is financed through a portion of the state's tax on the coal industry. The Partnership for Technology and Innovation of the state of Louisiana is funded through private industry, the state's economic development office, state utilities and infrastructure authorities, and private philanthropy. The nonprofit Science and Technology Foundation of the state of Maine has been given the legislative authority to raise funds through private donations, to market its own technical services to the private sector, to raise and invest capital, and to use investment earnings to finance its own operations.

Other countries have also developed institutional innovations to link technology to industrial growth. In Indonesia, significant emphasis has been placed on the creation of "business incubators," microfacilities with a trained management staff dedicated to providing the

> **An essential ingredient in virtually every successful partnership is concerted attention to strengthening human resources capacity.**

physical work space, shared facilities, and access to technical and business support services that are required to encourage small business growth. A national Incubator Steering Committee has been created and comprises senior government officials from the relevant agencies and private business executives. This dedication to creating early communication and joint stakeholding between government and private business is a fundamental strategy for ensuring the sustainability of what is intended to be a nationwide network of such incubators.

Human Resources

An essential ingredient in virtually every successful partnership is concerted attention to strengthening human resources capacity. In the U.S., human resources weaknesses have presented particular challenges to many state partnership strategies. The state of Montana, for example, ranks fourth in the nation in terms of secondary education graduation rates but 23rd in terms of college graduation and 47th out of 50 in terms of scientists/engineers per 1000 in population. The technology partnership is devoting considerable attention to addressing these human resources deficiencies. Indeed, its technology investments have begun to attract back to the state technically trained Montanans who had left the state for lack of job opportunities. Similarly, in Oregon in the 1980s, two of the largest R&D-based employers, Hewlett-Packard and Intel, had to import 90 percent of their senior scientists and 50 percent of their technical labor force from out of the state. Creating human resources capacity within the state was a critical element in the technology strategy.

Such problems are mirrored in many countries. In Thailand, for

example, severe shortages of
skilled personnel represent a critical roadblock to further development progress. Colleges and
universities graduate about 4,000
engineers per year, but current
demand is closer to 7,000 per
year. Trained technicians are in
even shorter supply. Similarly, in
Brazil in the mid-1980s, the
industrial expansion strategy was

> **Especially in pluralistic polit-
> ical systems, the relation-
> ships between public policy
> and private action are fluid.
> Science and technology
> initiatives require constant
> attention and must be
> resilient in the face of
> changes in political realities.**

engendered when internal technical capacity in the labor force inhibited both growth and the deepening of internal industries.

Key Questions for Future Analysis

However well-designed a partnership effort may be, history
teaches that the partnership process is never complete. Especially
in pluralistic political systems, the relationships between public
policy and private action are fluid. Science and technology initiatives require constant attention and must be resilient in the face
of changes in political realities. The policy process is not always
deliberate; it is opportunistic and hence best serves institutions
that are flexible and innovative. As one decision-maker in
Louisiana has observed in tracing the history of the Louisiana
Partnership, "Policy is made in the context of too many choices,
too few dollars, and too little information."[3]

Many of the partnership initiatives currently under way are
relatively young. Time is needed for data to emerge that will
raise and/or answer a set of key questions about the ultimate
utility of various partnership approaches. One critical question
will involve employment effects. Does targeted technology policy

created and implemented by public–private partnerships result in net job creation? Or are jobs simply displaced from one industry to another? Similarly, do such partnerships result in increases in investments in research and development? And can R&D productivity be shown to improve because of closer industry–university–public policy collaboration? In sum, the "vital signs" for technology partnership strategies relative to economic growth are underdeveloped at this time. Careful thought about how to measure net impact will be essential as partnerships mature and there is greater demand from both public and private sources of finance for partnerships to demonstrate results.

A Concluding Thought

Knowledge, and innovation derived from knowledge, will underpin future growth and prosperity around the world. Knowledge is now as much an economic factor as land, labor, or capital. Those who make the effort to integrate institutions and interests into cohesive strategies for technology-based economic growth will emerge as leaders in the next century. Those who do not may be left behind. Effective innovation—to improve economic performance today and to create economic progress tomorrow—requires a partnership not only among academic, government, and private institutions, but also between science and technology and economic policies. ∎

References

1. World Development Report 1994: Infrastructure for Development. The World Bank, Washington DC, 1994.

2. Paul M. Romer, "Implementing a National Technology Strategy with Self-Organizing Industry Boards," *Brookings Papers on Economic Activity, Microeconomics*, No. 2: 345–397. 1993.

3. J. Trent Williams, "Science and Technology Policy Making in Louisiana," *Science-Based Economic Development: Case Studies around the World*. Annals of the New York Academy of Sciences. **789:** 89–109, 1996.

Integrating the Military and Civilian Technology Base: A Long and Winding Road to—Where?

Based on remarks delivered at the New York Academy of Sciences on May 24, 1995.

STEPHEN M. DREZNER
Director, Critical Technologies Institute, RAND

D ual-use technology strategies are underpinned by two important, complementary rationales. On the one hand, dual-use strategies are seen as combining the best commercial technologies and capabilities with defense systems design expertise to produce necessary military technologies more quickly and at a lower cost than has characterized defense technology R&D in the past. On the other hand, such strategies are seen as reinforcing and strengthening the link between defense research and development expenditures and the nation's overall commercial performance and economic growth.

Attractive though both of these arguments are, each encounters difficulties on the road between the theory and the practice of integrating military and civilian technology development.

The Reality Driving Dual-Use Strategies

From the defense perspective, the current period of budget reductions clearly implies a decline in defense-specific research and development resources. But the demand for advanced technologies by the military—both for weapons systems improvements and for enhanced efficiency—will continue. Hence, defense R&D will need to become more involved in the global networks that currently dominate commercial enterprises. Reliance on the commercial sector will be essential to getting better products developed faster at lower cost. This link between the nation's defense and its commercial capability is not optional, it is mandatory.

Why is this so? In part, the answer can be seen in the composition of R&D resources in the United States relative to its global competitors. Viewed in terms of the distribution of defense and non-defense spending for R&D the U.S. is unique in the defense component of its R&D spending pattern, spending many times more than its industrial competitors. Viewed relative to economic activity, however, the picture changes (see Figure 1). U.S. total R&D investments resemble those of Japan and Germany, and the ratio of defense-related R&D resembles that of France and Britain. The R&D pattern of the U.S. is becoming similar to that of its economic competitors.

Moreover, the roles of the commercial and defense sector in the advancement of technology have also shifted. Decades ago, it

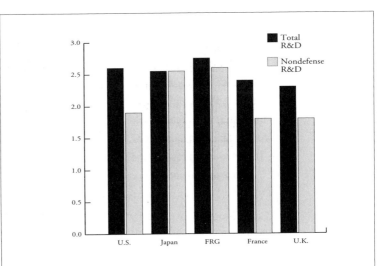

Figure 1. Proportion of non-defense R&D to total R&D as percentage of GDP.
Source: Critical Technologies Institute, 1995.

was defense R&D spending that led to technological advances. Today, in every technology area in which it is involved, the commercial sector is the leader. Defense technology is a leader only in those areas in which it is unique. To a large degree, commercial advances have out paced those of the defense sector because cumbersome, often needless, military specifications and cost accounting systems have so slowed the research and development process that it cannot keep pace with its commercial counterparts.

Three Questions and an Illustration

Against this background of change, three critical questions for dual-use technology must be answered. First, how can the

> **Reliance on military specifications for defense technology no longer makes any sense because the Department of Defense is no longer a technology leader in most areas. That leadership rests in the commercial sector.**

nation's defense system maintain its technological edge when the global marketplace can access the very same technology as the U.S. military? Second, can the military sector build its capabilities and systems around commercial technology? And, third, what is the link between a dual-use technology strategy and the economic competitiveness of the United States? The nature of and answers to these questions require constant vigilance because the nation's security depends on maintaining technological superiority. An example of the importance of the military–civilian technology interface will illustrate the importance of the issue.

The Global Positioning Satellite (GPS) system is used by both the military and civilian sectors for navigation. The 24 satellites act like a global lighthouse and are available to anyone who has the land receiver to use their signals to determine land, sea, or air location. The system has become predominantly commercial (see Figure 2), with the commercial market for receivers predicted to far out-pace that of the military by the year 2000. But, the military implications are clear. The wide availability of the receiver technology will increase any enemy threat to the U.S. military. The more sophisticated the missile system facing the U.S. military, the greater the marginal value the GPS system will provide the enemy. The more widely available the commercial capability, the more accessible those marginal values will be.

Clearly, then, rapid commercialization of a technology results in loss of military control of that technology. On the other hand,

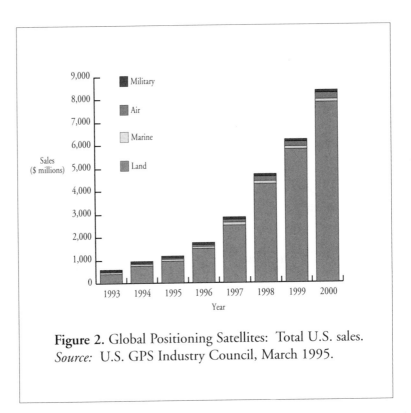

Figure 2. Global Positioning Satellites: Total U.S. sales. *Source:* U.S. GPS Industry Council, March 1995.

there are clear defense benefits to commercial availability. In Operation Desert Storm, for example, the U.S. military did not have sufficient receivers in stock to meet the operation's needs. The military GPS receiver weighed 17 lbs, cost $34,000, and required 18 months to build. Faced with a shortage, the military bought receivers from the commercial supplier. Each commercial receiver weighed 3 lbs and cost $1,300. There were differences, of course, in technical capacity, but the contrast is still striking. Parenthetically, by using best industry practices, the military has been able to reduce its unit cost to $1,200. The commercial unit now costs $800.

> The national economic implications of dual-use strategies are unclear. There are no comprehensive, reliable measures for isolating the effects of research and development investments in overall economic growth.

Maintaining a Technological Edge

The commercial and military sectors will share technology under a dual-use strategy. The technological edge will not come from controlling technology but from changing the parameters of the interface between technology and military needs. There are three dimensions to that change.

First, the military must try to ensure that U.S. commercial technology gets to market faster. In the past, complex software systems might take ten years to design. But the world did not stand still, and the software would be obsolete by the time the weapons entered production. Thus military cooperation with commercial developers is critical.

Second, the military will need to develop ways to better utilize technology. New systems for integrating technological advances into military needs, as well as improved approaches to training, are required. Finally, the military will need to use technology to improve the process involved in national defense. Information systems, modeling, and simulation can greatly benefit from commercial technological advances.

Military Requirements versus Commercial Resources

Goal conflicts in dual-use strategies are inevitable. The cost versus performance ratios that are acceptable in thinking about promoting competitive advantage may not be so acceptable when the goal is to defend the country. Merging the defense and commercial technology base of the United States will raise such conflicts.

If weapons systems are going to be based on commercial technology, then how will government influence the R&D underlying that technology so that military requirements will be met? Resolving this conflict will require a top-to-bottom transformation of the current military system for relating to the commercial sector. The military will need to begin to proactively design systems that can use commercial components. In turn, this will require not simply a change in procurement systems, but changes in the entire culture of the Department of Defense. The defense sector will no longer have the luxury of controlling its own industrial infrastructure. A commercial technological interface will not be a preference in weapons design, it will have to be a requirement. In turn, this will mean that the military will need to be concerned about the state of the nation's technological capacity not only in research, but also in product development and, ultimately, production.

> There are very real goal conflicts in dual-use technology strategies. The cost versus performance ratios acceptable in thinking about promoting competitive advantage may not be so acceptable when the goal is to defend the country.

It will take time and resources to change the culture of defense. Productivity in the military will change in complex ways, and the associated learning curves will have hidden costs. Fortunately, the leadership of the U.S. military is committed to change, and is even working with such corporations as Motorola and General Motors to understand the nature and process of organizational change so that the defense process will benefit from organizational experiences elsewhere.

Dual-Use and National Economic Competitiveness

Although the Department of Defense's share in total U.S. R&D

has been declining, national security R&D continues to play a very large role in the total R&D budgets. In 1996, budgets for military basic research, applied research, and development, total over $37 million out of a total national R&D price tag of $70 million. How that money is spent, however, tells the tale of its importance. Of the nation's 1996 $41.6 million budget for technology development, $32.6 million, or 78 percent, will emanate from the defense sector.

Hence, the link between military expenditures and the nation's technological (and thus economic) superiority appears to be fundamental. However, that relationship is much more difficult to measure than to posit. Trying to determine with any confidence the fraction of total factor productivity (or any other output measure) attributable to technology has proved to be daunting. Moreover, the technological factor pales in comparison with other elements, such as management and marketing, in much industrial analysis.

The relative productivity of government-funded versus privately funded R&D has also been questioned. In the view of some analysts, defense R&D spending is critical to maintaining America's position in world technology. To others, the investment returns on government-sponsored R&D are so much lower than the returns to private R&D investments, that the economic rationale for maintaining defense capacity is undermined.

Dual-Use Within Broader Technology Policy

A dual-use strategy is only one, albeit crucial, element in a broader set of technology policies. Dual-use policy focuses on a somewhat narrow set of industries (e.g., aerospace, electronics, and information processing), and spin-offs that benefit the nation's overall economic health may be limited. U.S. technolo-

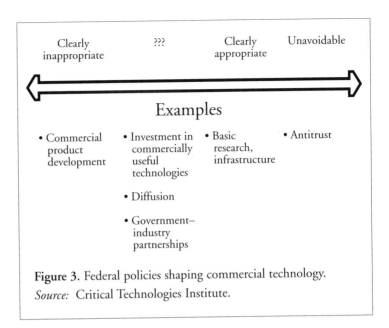

Figure 3. Federal policies shaping commercial technology. *Source:* Critical Technologies Institute.

gy strategy obviously needs a more comprehensive approach.

As in many other debates, however, the critical problem in crafting such a multi-dimensional approach to linking technology to economic growth turns on the role of government policies and programs. A variety of federal policies shape commercial technology. These policies can be arrayed along a spectrum as displayed in Figure 3. In some areas, such as antitrust or intellectual property standards, the federal role is not only appropriate, it is unavoidable. On the other end of the spectrum (e.g., commercial markets), federal involvement is clearly inappropriate.

As always, it is the grey area in the middle that sparks debate. Support for basic research infrastructure for research on pre-competitive generic technologies has recently been seen as an appropriate area for federal support. But such areas of agreement are few. Commercial investments, technology diffusion, and government–industry partnerships for technology development all are under heavy, often impassioned debate. The ultimate result of that debate will determine the nature of federal technology policy well into the future. ■

IV

Ensuring Resources

Money, as the *Cabaret* song goes, makes the world go 'round. Significant resource commitments are required in both the public and private sectors if research and development are to continue to provide the level of innovation that is needed to fuel future prosperity. But resource commitments are required in many sectors; science and technology do not stand alone in the investment line. And that line is long. How to make resource decisions, how to balance trade-offs, how to ensure that resources committed are producing results efficiently—are all complex problems for both public budgetary authorities and private corporate executives. Getting public and private resource investment decisions right is important; the resource stakes are high. In the United States, over $170 billion is invested in research and development. In countries such as Japan and Germany, the absolute resource levels allocated for R&D may be less, but the percentage of gross domestic product dedicated to R&D exceeds that of the United States. The commentaries that follow illuminate the varied, and often controversial, dimensions involved in making those resource decisions.

The Future of the National Laboratories: Navigating the Route to Renewal

Based on remarks delivered to the New York Academy of Sciences, in collaboration with Business Executives for National Security, on March 14, 1996.

ROBERT W. GALVIN

Chairman, Executive Committee

Motorola, Inc.

Former Chairman

Task Force on Alternative Futures for the

Department of Energy's National Laboratories

F or decades, one of the central elements of the nation's security system has been the system of national R&D laboratories attached to the United States Department of Energy (DOE). These institutions have become premier centers of research and technology development. They are at the cutting edge of discovery across scientific disciplines. But, with the end of the cold war, times have changed. Whereas in the past much of the DOE labs' attention was focused on critical goals of national security and military preparedness, their future promises to be much more diffuse and complex, with a shift in emphasis toward competing in the global economy.

Triple Roles for the Laboratories

First, the national preparedness mandate will continue, albeit at a different level and for a wider range of problems. America must continue to be a strong and prepared nation in the conventional sense, as well as relative to new threats such as combatting drug trafficking and international terrorism. The research capacity of the laboratories will provide a core capability in evolving future readiness.

> The future promises to be diffuse and complex, with a shift in national laboratory emphasis toward the global economy.

Second, the legacy of the past remains with us physically, however much it has unraveled politically. The threat of global nuclear war may have faded with the demise of the Communist system, but the stockpile of weapons technology that made the threat so very real is still to be found throughout the world. Managing and/or dismantling that capacity will require continued commitment of resources, perhaps as much as a quarter of laboratory resources over the next several years.

Third, and even more recent, however, is the laboratories' role in forging a linkage between the R&D capacity of the national laboratory system and the commercial marketplace. The relationship between the nation's federal R&D capacity, represented by the national laboratories, and the competitiveness of its private sector is one that will need to be carefully planned and executed.

In turn, binding all three of these roles into a unified laboratory mandate and operating system will require new management techniques. The Task Force on Alternative Futures for the Department of Energy National Laboratories spent considerable effort in 1994 on examining the needed adjustments in the objectives and operations of the laboratories, and, especially, the implications of these adjustments for laboratory governance. Although Task Force recommendations have not been fully implemented, they have led to a deeper discussion of what must change and why.

Who Needs to Care about Change?

The critical audience for the core recommendations is really the United States Congress. Although the

> **Congressional micro-management of national R&D and technology development is not a healthy role.**

Department of Energy has responsibility for the laboratories, those who need to be convinced of the need for change sit on Capitol Hill.

There is a debilitating practice by Congress to ordain micro-management of the institutions, to specify in great detail what institutions should do and how they should do it. I would assert that this is not a healthy congressional role in general, and certainly is not optimal in the area of R&D and technology development. The national laboratory system is staffed with some of the world's best scientists and engineers, with relationships to a network of the best academic and private institutions in the country. It is inconceivable that congressional micromanagement would add value to their programmatic decision-making.

"Corporatizing" the National Laboratory Asset

One major recommendation of the Alternative Futures Task Force was that the national laboratories be corporatized. What are the implications of such a move? A few definitions will help to clarify the distinctions between two approaches—privatization and corporatization.

"Privatization," a much discussed yet generally poorly understood term, means that the laboratories would be owned by private, commercial shareholders, investors, or institutions. The equity present in the laboratories would be acquired by private investors. Although this may ultimately happen, it is not likely to fit the nation's current capacities. No large corporate entity presently exists

> Embedded in a corporatization strategy is trust—trust in government that an apolitical, private group can manage a key element of the nation's technological capacity.

that is big enough to be able to purchase the value represented by the national laboratory asset. This may change, and I will return to that possibility later.

The alternative is to "corporatize" the laboratories. In this strategy, the laboratories would continue to be owned by the United States government. The citizens of America would, in essence, be the shareholders. The government would provide all of the laboratory assets to a new corporation.

The governing board of the corporation would comprise private professionals, from business or science, who would serve for no compensation. There is no question in my mind that the most senior, experienced business executives and scientists in the nation would volunteer to serve in this fashion.

The executive and legislative branches of government would then entrust stewardship of the laboratories to the corporate board. The board would name and oversee the management of the corporation. Embedded in the word "entrust," of course, is the key assumption behind the strategy—trust. The political process would ultimately need to trust that an apolitical, private group of individuals can manage a key element of the nation's technological capacity. Trust is not a comfortable concept in many halls of government, which is perhaps why the corporatization option has made such slow progress.

What would be the benefits from corporatization? Most importantly, the ability to make and implement decisions at the laboratory level would reduce the administrative overhead associated with managing the institutions. Greater control of laboratories over their own management could result in up to 40 percent cost reduction—not 4 percent, 40 percent. That may seen unbelievable, but management studies in the private sector indicate that such savings (or, conversely, equivalent increases in productivity) result from organizational changes that bring management control closer to operations and private initiatives of quality management practices.

> Someday, the national laboratories will be attractive corporate acquisition candidates.

Why Not Privatize?

If such savings could result from corporatization, why not just privatize the system? This is simply not financially possible at this time. Each of the laboratories is engaged in between 0.5 and 1.5 billion dollars of R&D per year. The capital cost to purchase each of the laboratories would be approximately the same. There is simply no U.S. technology-based company rich enough to afford to buy one of the laboratories at that price.

It might be possible for a consortium of companies to access such a level of capital, but then the question would be efficiency. Even a consortium of companies would not want to purchase the laboratories as they currently exist because they have not been allowed to operate at a level of efficiency that would allow them to blend into a corporate structure.

Nevertheless, privatization is certainly imaginable in the future, particularly if the management changes that corporatization would allow were in force. Some U.S. technology corporations are growing very rapidly. I believe that a day will come when certain of them will have annual revenues in the upper hundreds of billions of dollars. The corporate engineering budgets, which normally constitute 10 percent of corporate expenditures, will then be larger than the budgets of the laboratories. At that point, some national laboratories will be attractive takeover targets.

National Laboratory Future Priority: Energy

Given that, as already noted, some 25 percent of national laboratory work is devoted to national defense work, where should the other 75 percent go? The Alternative Futures Task Force recommended that the focus be on energy. We are on the eve of possessing the means to have low-cost, clean, affordable energy generation in place around the world. That energy will allow rapid increases in productivity, which will, in turn, expand incomes and investment and fuel global economic growth. More widespread sources of energy are critical for future growth.

At Motorola, we look increasingly at the opportunities around the world as an algorithm of population rather than of gross national product. Soon, the world will support six billion people. The U.S. will be about five percent of that total, a robust, dynamic five percent to be sure, but, in terms of the profile of our business, 95 percent will be outside of the U.S. But, we cannot enter and serve those markets without energy. It is critical for local development and for the increasing incomes that will allow healthy economic relationships.

In many ways, it is energy that will allow the kinds of investments necessary to take advantage of the great opportunities that science and technology now can place before global society. Continued progress will require new energy capacity. Developing the technology that will create that capacity could be the major challenge of the national laboratory system. If we succeed, it will mean scores of new industries all over the world, and a new era of global prosperity. ■

Choices Amidst Change: S&T Resource Priorities for U.S. Global Leadership

Based on remarks delivered at the New York Academy of Sciences on June 12, 1995.

JOHN H. GIBBONS

Assistant to the President of the United States for Science and Technology

n many aspects of the role of science and technology in the nation's future, it is possible to find broad agreement across a spectrum of policy views. Congress, the Administration, and most commentators agree that the United States has embarked on an era of great change; that science and technology bring economic rewards to society; that discovery is unpredictable and research essential; that failure to maintain a robust investment in research and development will cripple U.S. competitiveness; and, that science and technology are critical to the nation's future. Indeed, Vannevar Bush, author of a seminal work on U.S. science policy, made the very same observations fifty years ago.

> **The question is not whether science and technology are good things. The question is how to structure a budget that yields these good things without robbing Peter to pay Paul.**

But such agreement can be misleading because it can mask more fundamental questions. The answers to those questions do, indeed, engender deep debate. The question is not whether science and technology are good things, but rather how the nation structures a budget that provides all of these good things, and does so without leaving essential programs bereft of resources simply to fund another equally essential program? Moreover, how can such an effort be accomplished in ways that generate support among the American people? Without the public's willingness to make the kinds of public resource commitments that S&T investments require, the government's role would diminish greatly.

The Tapestry of Science Funding

The current trend in congressional budget proposals argues for deep cuts in the nation's science and technology investments. There are those who say that such cuts will not damage the fundamental fabric of science that has supported this country for three centuries. They are mistaken.

The current system of R&D funding is the product of trial and error and painstaking reassessment dating back to the birth of the Republic, and it stems from three overarching principles. First, funding for science and technology serves broad public goals as well as specialized commercial or military interests. Second, this funding yields a better return on the dollar than nearly any other federal investment. Third, science and technology are best done across the scope of the federal enterprise, woven into the public programs of a wide range of departments and agencies.

The result of these principles and the associated funding system is a rich tapestry of research and development that has been responsible for as much as half of the nation's economic productivity since World War II. This tapestry is a remarkably sturdy and resilient piece of art. It responds to the threat of new and emerging diseases such as the Hanta and Ebola viruses. It supports a space program that can put our minds where our feet can never go. It prepares American children for the high-technology, global economy of the future. It has contributed to a world at peace, and yet continues to guard the nation's security. It protects the air and water, and puts food on the world's tables.

> **Current federal funding patterns have resulted in a rich tapestry of R&D activity that has accounted for as much as half of U.S. economic productivity since World War II. Congress is hacking at that tapestry with garden shears.**

But however strong and carefully crafted the tapestry is, it is not indestructible. If a thread here or there is pulled, perhaps the overall effect is not noticeable. But if a third of the threads are pulled, the intricate fabric may become simply a mass of tangled yarn. That is the risk the nation's science and technology enterprise now faces. By deeply cutting the government role, Congress is hacking at the fabric of science and technology with garden shears.

Programs At Risk: Some Examples

Three examples will serve to illustrate the degree to which unwise budget decisions can endanger the nation's fundamental science and technology capabilities.

America's weather forecasting capacity is absolutely dependent upon the satellite assets of the National Oceanic and Atmospheric Administration. Maintaining the current system and replacing older satellites as they near the end of their productive lives, for example, is essential for the purpose of providing early warning of Atlantic hurricanes or Pacific storms. These capabilities not only protect property, but also save lives. Current congressional budget

> **Much of America's native optimism rests on the belief that advances in science and technology lead to a better life.**

proposals would eliminate fully half of that capability.

Similarly, congressional budget proposals have called for $1 billion in cuts to the National Institutes of Health (NIH). The result would be a potential reduction by one-third in the number of research projects that the NIH funds every year. Which bright young researcher will be left behind? Which breakthrough in curing a killer disease will remain unknown? What huge long-term price will the nation then pay for a short-term budget decision?

Basic science will also face a crisis. Budget proposals from the House of Representatives would result in the National Science Foundation's cutting 4,000 research awards and funding 11,000 fewer scientists, engineers, and students than did the fiscal 1996 and 1997 budgets proposed by the Administration. At the Department of Energy (DOE), congressional budget proposals would pare 35 percent from basic science. As many as ten facilities operated by the DOE for collaborative work with universities and industry would have to be closed, resulting in the loss of 8,000 jobs. These basic research facilities have given America a cutting-edge capacity not only to lead the world in scientific discoveries, but also to address practical needs such as the search for energy sources with less pollution, better computers and electrical appliances, and even a longer-lasting light bulb.

Support for High Technology

But it is not just the R in R&D that is under threat. Technology programs requiring development support may face even greater peril.

A great deal of America's native optimism rests on the belief that advances in technology lead to a better life. America is the world leader, or shares the lead, in 27 technology areas that are critical to national and economic security. Yet that lead is fragile or decreasing.

Of those 27 areas, Europe is dead-even or only slightly behind in 25. Japan is tied or slightly behind the U.S. in 17 of the 25, and closing fast in 5 others. Maintaining leadership, or even parity, means continuing to move forward.

> **America's sturdy tradition of government collaborating with industry to develop new technologies with large public payoffs has brought widespread economic growth.**

The irony is that, just as the United States is considering abandoning its historic commitment to science and technology, the nation's economic competitors are exploiting their own commitments. With its $7 trillion economy, the United States still leads the world in total dollars spent on R&D. Yet, Japan and Germany far surpass U.S. levels of spending on nondefense R&D as a percent of Gross Domestic Product. The U.S. cannot ignore the fact that the world is filled with fast-moving, highly capable competitors, aided by governments that work with them to develop advanced technologies.

America has a sturdy tradition of governmental collaboration with industry to develop better technologies with large public payoffs. That tradition has contributed to widespread economic growth, better health and longer lives, and a strong national defense. In today's world, that tradition is essential to ensure that America's businesses and workers are competing on an even playing field.

The Administration's technology partnerships with private industry, universities and colleges, and state and local governments, embody that tradition of collaboration. These technology partnerships are not "political pork," special favors for politically connected firms or powerful industries, they are investments that enrich the economy. In the partnership program, independent experts choose projects strictly on the basis of merit with industry sharing the cost. That cost-sharing provides the program's market test, ensuring that the private sector, not government, is placing bets on the likely winners.

> **The nation faces a new and uncertain future where the road maps to success are unknown. Knowledge is the only clear advantage in such a time. Knowledge is the key to the future.**

Cutting the Budget and Protecting the Tapestry

There is no question that the federal budget will be scaled back as a result of ongoing budget negotiations. But there is a right way and a wrong way to accomplish that goal. Just as fixing or re-cutting a tapestry requires a master weaver, so it is essential that adjustments in the nation's R&D investments be made carefully. The Clinton Administration has been engaged in that enterprise for the last three budget years, resulting in more than $600 billion in deficit reduction. Despite deep cuts in virtually every other domestic discretionary account, research funding has actually risen modestly, a signal of the Administration's commitment to science and technology as the engine of growth in jobs, the economy, and the quality of life. Indeed, basic research has received the greatest percentage of that increase.

Continuing the budget reduction process will require careful examination of the reverberations of any particular budget cut on the interdependent web of science and technology across the nation. Indeed, federal S&T programs are increasingly interdependent across agencies and across budget line-items. Significant cuts to one program can have substantial effects on other national initiatives. These inter-relationships must be carefully understood, and budget reductions must be cautiously approached. A budget-cutting philosophy that uses "fast and deep" as its two central criteria is the opposite of the approach needed.

The United States faces a new and uncertain future, where roadmaps to success are unknown. Knowledge is the only clear advantage in such a time. Knowledge is the key to the future, and only investments in science and technology can forge that key. Lewis Thomas has said that the greatest discovery of this century will have been the discovery of the extent of our ignorance. We must not let this ignorance become public policy. ■

Science, Technology, and the 104th Congress: An Interim Report

Based on remarks delivered at the New York Academy of Sciences on June 2, 1995. Since January 1997, Mr. Walker has served as president of the Wexler Group in Washington D.C.

ROBERT S. WALKER

Chair, Science Committee, U.S. House of Representatives

S
cience and technology provide an essential foundation for the future of this nation. Just as constraints on resources require choices to be made in virtually every other area of public life, choices must also be made in our allocation of government resources to science and technology programs. Because cleverness, which in this case means the ability to manipulate information, will create the wealth of the future, we must carefully determine those priorities in science that will lead to the development of new knowledge.

S&T Policy within Four Revolutions

The science policy decisions that are now being made in the 104th

Congress exist within the context of an era that is revolutionary in four respects. In terms of political organization, the United States, like many other nations around the globe, is moving away from centralized bureaucracy and toward smaller and more local government. In terms of economics, the revolution is being defined within a framework of the evolution of highly industrialized economies into knowledge-based economies, and the transformation of national centers of economic power into global ones. As a consequence, culture, too, is undergoing change and analysis.

> An investment in basic research is an investment in the future. Government is the only agency which can make an investment in basic research without having an end product as a goal.

Finally, science and technology are affecting every facet of life, with scientific research and technological innovation changing how we communicate, learn, travel, and even, given the ever-present fax and e-mail systems, how we make policy.

Within this revolutionary context, three general principles underlie current congressional decisions about the substance of S&T policy. First, government must be transformed to fit modern realities and future opportunities. As it is presently constituted, government has been defined by an age that is disappearing in the face of global economic and political change. But since change also brings new opportunities, government must adapt, not only to shed the old, but also to take advantage of the new. Second, the nation cannot spend money that belongs to future generations. Living today on the money that will be earned tomorrow by our children and grandchildren is both morally and economically wrong. Third, what worked in the past is not necessarily adequate for the future. The performance record of a program in an indus-

trial economy does not necessarily argue for its continuation in a
knowledge-based economy.

Looking Inside the National Budget
The budget offered by Congress
is premised on re-evaluation,
restructuring, and reform. The

> **If fiscal 1995 is taken as the
> base, the proposed budget
> for fiscal 1996 will actually
> increase those accounts
> allocated to basic research.**

budget is not simply about cutting programs; it is also about dramatically restructuring government. Indeed, while balancing the
budget over the next seven years, the congressional proposal would
spend 11.7 trillion dollars in federal programs, 2.2 trillion dollars
more than has been spent in the last seven years. Congress believes,
therefore, that it is possible both to balance the budget and to have
more spending.

To do so, however, requires restructuring. As a result, proposals
are under consideration to eliminate three cabinet departments,
283 separate programs, 14 federal agencies, and 68 federal commissions. This is the beginning of the process of rethinking government and reforming outmoded institutions and programs.

Science and Technology Budget Proposals
Over the next five years, the congressional proposal is to spend 111
billion dollars in civilian research and development. While the fiscal 1996 budget is smaller than that of fiscal 1995, it is still 18
percent higher than that at the beginning of the 1980s, before the
dramatic budget increases of the late 1980s and early 1990s.

If 1995 is taken as the base line, congressional proposals for budget allocations to basic research show, in fact, a nominal increase.
This reflects a conscious decision on the part of Congress to reduce
prior emphasis on corporate technology-support programs and to

> **The space station is an extremely important part of a very complex relationship between the United States and Russia.**

increase investments in basic research. While, of course, the line between basic research and technological applications is not always broad and clear, basic research has clearly been chosen as a priority because it is truly an investment in the future. Basic research is the one place government can allocate money without depending on an end product in the marketplace, thereby beginning to build the essential foundations for future discoveries and development.

A second priority that has underpinned science and technology budget decisions has resulted from the belief that federal science policy over the last twenty years has been increasingly jeopardized by political considerations. Rather than allowing peer review science to exercise basic judgments about resource allocations, the political process has been used to determine which programs live and which do not. In essence, then, the Budget Committee's structural priority is to restore the independence of such fundamental institutions as the National Science Foundation and to make them more immune from the programming pressure originating on Capitol Hill.

A Case in Point: NASA and the Space Station

Our approach to structuring the budget for the National Aeronautics and Space Administration (NASA) illustrates these several points. Funding for the space station has been preserved, a controversial decision that nevertheless served to re-establish the two priorities just discussed. First, Congress recognized that the space station is an essential part of human endeavor in space, which itself is critical to basic research. The space shuttle system will require a

> **The Advanced Technology Program has begun to eat away at the core mission of the National Institute of Standards and Technology, endangering one of the most valuable assets of the nation.**

space station sometime in the next century. Second, and perhaps more importantly, space station funding turns not only on science and technology issues, but also on the geopolitical and economic revolutions mentioned earlier. Congress and President Clinton agree that continuation of our commitment to the space station is an extremely important part of a very complex relationship between the United States and Russia. Discontinuation of the program would have endangered this country and the government of Russian President Boris Yeltsin.

Because resources are limited, however, this choice required other sacrifices. Preserving the space station will require reducing other areas within the space budget and extracting considerable savings from management reforms within NASA itself.

The Controversy Over Support for Advanced Technology

Perhaps the greatest controversy between the President and Congress has been over congressional decisions to reduce or eliminate the Advanced Technology Program (ATP) of the National Institute of Standards and Technology (NIST). The ATP is intended to provide grants to private corporations for technology development. The problem is that the ATP has begun to eat away at the core mission of the NIST, which is to set scientific and technological standards; thus, one of the nation's most valuable assets is endangered. There appears to be a connection between the recipients of ATP grants and donations to political accounts in Washington, a state of affairs which, in my view, hampers reasonable choices among science priorities.

> **Congressional decisions about science and technology must rely on a system of peer review and not politics.**

Similarly, the Department of Energy carries out a variety of high-cost research projects in technology which have led to few commercial applications. Government programs that try to use public funds to pick commercial technology winners and losers do not have an impressive record of success. The process of government grant-making is simply too removed from the ultimate marketplace to be sensitive to whether one particular innovation is better than another and deserving of public subsidy. There is nothing particularly new about these observations, either. Recently, a constituent sent a letter that he had found in his attic written to his father in 1917 from the National Advisory Committee for Aeronautics. In part, the letter read:

> *There seem to be two fairly distinct questions with regard to the value of any parachute device. First, whether it might be developed to the point of usefulness for aviators under normal conditions, and second, whether, under any conditions, it may be considered of military value. The present attitude of military authorities seems to be a negative answer to the latter question. It does not follow, however, that such devices may not have some possibilities, especially in the future commercial or industrial uses of aircraft. With such a possibility and view, development should be encouraged, at least as representing a possibility of future usefulness.*

Of course, the national committee had the author's conclusion exactly backward. There is no military pilot today who climbs into an aircraft without a parachute, and no commercial passenger who dons a parachute before takeoff. The point here is that bureaucratic and political personnel are unlikely to produce the most astute scientific and technical analyses. Peer review, not politics, should underpin scientific decisions and scientific resource allocation.

The Disjunction between the Budgetary Process and Technological Innovation

Another reason why direct federal grant support for the development of advanced technology encounters

> **Many consumer electronics products become obsolete in eighteen months, about the time it takes for Congress to complete the federal budget process.**

problems is the disjunction between the federal budgetary process and the pace of innovation. The time lag between the first estimate of the federal budget for a fiscal year and the final passage of an appropriations bill in Congress is now about 18 months. Yet in 18 months an entire generation of consumer electronic products comes and goes. Hence, making judgements about allocating federal funds to particular technologies risks making those technologies obsolete by the time the funds are actually available.

Moreover, science is not just a line item in a program budget. A wiser strategy is to look at science and technology investments not in terms of particular products, but in terms of the nation's overall R&D expenditures. Government budgetary roles should be combined with other public policy levers that will create private investment incentives. If the government focuses direct resource allocations on basic research, it can promote the application of that research not by picking winners and losers itself, but rather by using tax policy, particularly linking the R&D tax credit for corporations to their relationships with universities, motivating commercial decision-makers to invest in those applications. As a consequence, commercial decisions are much more accurate and timely when they are made in the marketplace. Technology transfer, then, would be decoupled from government programs and become a product of commercial decisions motivated by improved tax laws.

The Department of Science Option

As government is restructured and reformed, reorganization of government agencies will also be necessary. In the cases where the elimination of agencies leaves behind critical S&T functions, these latter will need to be regathered or rehoused so that the best efforts are not lost with the elimination of outmoded programs. If, for example, the Department of Energy is eliminated, the functions of the weather research program would clearly need to be retained. There are similar situations in other public agencies whose overall mandate is outdated, however meritorious a set of particular subcomponents.

Possibly these pieces could be restructured into a Department of Science. Clearly, such a department would not encompass all federally supported science. Rather, it would provide a means for rationalizing what would be a disparate array of scientific functions that would need to survive the pruning of government structures.

The Imperative of Reform

Adjusting to a changed global economy, a new geopolitical world, and a more rapid pace of innovation, will require a rethinking of priorities throughout the federal government. Change is difficult in general and no less difficult in science and technology policy making. But change is happening around the globe. The critical challenge for government is to create a role that is focused on investing in basic science which in turn will lead to prosperity for the generations that follow. ■

Civilian Technology for Economic Growth: The Changing Face of Federal R&D

Based on remarks delivered at the New York Academy of Sciences on May 24, 1995. Dr. Prabhakar is now chief technology officer and senior vice president at Raychem Corporation in Menlo Park, California.

ARATI PRABHAKAR

Director, National Institute of Standards and Technology

T echnology is the most dynamic, non-linear tool that can most fundamentally change the ground rules of economic competitiveness. Education and training, ability to market, and capital all matter. But technology is the only tool that can change the ground rules in all of these other areas.

The Evolution of Technology in Private Industry

To understand the economic environment within which technology functions, one must turn to the evolution of private industry. Often the public thinks of technology in terms of the ability to walk on

> **Technology is the most nonlinear tool that can effect the most fundamental changes in the ground rules of economic competitiveness.**

the moon or conquer polio. These were and are noble roles for technology. But technology has a powerful role to play in ensuring the fundamental health of the nation's private industrial base.

After World War II, American industry represented the majority of the world's manufacturing capacity. Innovation from the United States drove the nature of the world market. In just a few decades, that picture has changed radically. The global economy is fiercely competitive. In many industries, U.S. leadership has ebbed; in others it has remained dominant. But, even when an industry is king of the mountain at the moment, it must constantly run harder and faster just to keep up with competitors. In this way, the marketplace is truly working because it is driving the efficiencies and innovations that lead to new products and rising standards of living.

The premium now placed on innovation has also changed the importance of R&D in private industry. There was a time when an investment in R&D could be paid over many years, even decades, by profitable manufacturing and sales. Today, product life cycles are measured in months, and a week's delay in product roll-out can spell the difference between success and failure in some areas. These more rapid cycles are, in some ways, driven by technology and, in some ways, they have technology as their product.

The dominant theme in industry, however, is that, with technology in such a fundamental role, the process of innovation must be efficient, and the link between innovation and product must be effective. In driving toward these goals, private industry has significantly restructured its R&D activity. The days of centralized, large-scale, basic scientific research in corporate America have passed. Corporate R&D has become more decentralized, more market-focused, more closely linked to immediate, continuous needs to meet global competition.

. . . and Lessons from the History of Federal Programs

These realities also call for a re-examination and re-structuring of the federal government's role. At the end of World War II, basic government investments in science and

> In today's economic climate, even if you are the king of the mountain at the moment, life in industry is one of always running harder and faster to stay caught up with innovation.

engineering were seen as critical. The result was a world-class university research system that continues to lead the globe in basic research. We also invested very heavily in technologies for national security. This investment in national security R&D was an important factor in shifting the balance of power and ending the Cold War.

It is important to recognize that the success of earlier federal investments took time. The results were not seen in the space of one congressional session or even one presidency. Success required staying-power, a willingness to invest over long periods before palpable results could be seen. After all, the fundamental reason for government programs in these areas is the long-term, broad-based nature of these investments.

A second lesson from past success is the importance of a carefully designed relationship between the government and the technology community. The relationship between the federal government and the university system, or between the Pentagon and its technology development contractors was carefully tailored to the job at hand.

> There was a time when R&D investments could pay off over many years. Today, product cycles are measured in months.

. . . Calls for the Evolution of the Government's Technology Role

The changing environment of U.S. economic competitiveness, and the fundamental importance of technology within that environment and the long-term nature of technology investments, call for an evolution of the federal government's approach to its technology role. There is a need to move our R&D investments in a direction that will be more responsive to the needs of American society, and to do so in partnership with the critical actor in future prosperity—private industry.

The shift that is underway now enables a direct link between federal R&D investments and the economy. This is happening throughout the government, including, but far from limited to, the National Institute of Standards and Technology (NIST). The evolution is not random; it is purposeful, seeking to create the kinds of relationships with private industry that will lead to long-term economic impacts in a very direct fashion.

Resources are limited. NIST represents just one percent of all federal R&D investments, which themselves are smaller in dollar terms than private sector R&D. But NIST's technology programs are an essential bridge between the long-term investments and the frantic pace of the market.

NIST Programs

Measurement Standards

Created in 1901, the then National Bureau of Standards encompassed a series of laboratories that developed engineering and measurement standards to accompany rapid industrialization. That measurement infrastructure was essential. At the turn of the last century,

New York City alone used three standard measures for a foot! Obviously, confusion was rampant, and the spread of industrial enterprise was inhibited.

The standards laboratories changed all of that. Today, the

> **Unified standards are essential. At the turn of the century, New York City had three different standard measures of a foot!**

measurement standards role of NIST laboratories remains fundamental. The measurement issues, of course, are now about line width on a semiconductor, not the radius of a buggy wheel, but the commitment of NIST to its maintaining the measurement infrastructure is unchanged.

In the last two years, the budget for the standards laboratories has actually risen significantly for the first time in decades. NIST is now in a much better position to ensure that its responsibilities for the measurement infrastructure will be able to keep pace with the emergence of whole new industries and needs in areas such as biotechnology and information technology.

New Technology Partnerships

In addition and, in part, because of NIST's historically close relationships with industry, a series of technology initiatives has also been undertaken. Two examples will illustrate how the government's role is evolving within the new economic environment.

The Advanced Technology Program (ATP) funds projects in companies to undertake important enabling technology investments that have long-term economic importance but are longer-term and of higher-risk than companies will fund on their own. Competition is rigorous. Costs are shared with the companies. Awards are made on the basis of recommendations from experts in the business and technical community. Competition for the grants is keen: Since 1990, the ATP program has received about 1500 proposals and has made 177 awards.

A second program is the Manufacturing Extension Partnership, which provides small manufacturers with access to technology information and expertise. About half of America's manufacturing capacity is in firms that employ 500 or fewer people. These firms do not have their own R&D capacity, have seldom had to compete internationally, and few have access to new business practices and technologies that would improve efficiency and global competitiveness. The program tries to break down these barriers with information and technical expertise. Costs in this program are shared with state and local government, and assistance takes place not inside the Washington Beltway, but on the shop floor.

Government Versus the Market: A False Dichotomy

Some discussion has taken place in Congress regarding the relative merits of directly supporting technology development and diffusion compared to using tax credits to accomplish the same objectives. The theory is that, given sufficient tax credits, American corporations themselves will make the investments in basic research, university, and small-business partnerships, and in high-risk technology development, and will do so with greater efficiency and more attention to market dictates than will government programs.

A tax-policy-only approach, however, is flawed. If perfect tax policy and perfect regulatory policy could be developed, this still would not resolve the central problem. American companies today must cope with an increasingly fierce competitive environment. The pace of technological change will continue to accelerate. Competition and rapidity are forcing private industry to narrow, not expand, their R&D investments. We have had R&D tax credits for several years, and while this can be a useful financial tool for some companies, there is no evidence that such a policy led to significant changes in corporate R&D investment patterns. More of the same is simply not enough.

> **The fundamental reason for the government's past investments in basic science and technologies for national security is the long-term and broad-based nature of those investments. Such investments do not yield palpable results in very short periods.**

Back in Washington, we are living in a world in which there seems to be very little understanding of the kinds of forces that drive technology in the marketplace, and the kinds of dramatic changes that have taken place in the last few decades. We are still discussing science policies in a way that may have been appropriate twenty or thirty years ago. People I speak with in industry cannot believe that Senate and House budget resolutions have proposed eliminating the Department of Commerce, including important programs like our Advanced Technology Program and our Manufacturing Extension Partnership. One proposal would eliminate NIST laboratory efforts, which would be privatized—whatever that means. I don't think that there has been sufficient awareness that programs at NIST have been explicitly designed to recognize the magic of the marketplace and to really reduce barriers to enable industry to succeed in the marketplace.

There are a host of very real issues about the appropriate role of government in funding science and technology. Those questions should be asked, but they ought to be asked about the entire 70 billion-dollar investment. I am convinced that if we ask those questions, and look fairly at the kinds of investments that are being made and the needs of our technology systems, that these types of partnership programs will be a part of a future federal technology investment, even in a period of reduced budgets. ■

Will the U.S. Wellspring of Technology Dry Up?

Based on remarks delivered to the New York Academy of Sciences on March 10, 1996.

WILLIAM J. SPENCER

President and Chief Executive Officer

SEMATECH

T he last decade has witnessed significant change in the organization and conduct of industrial research and development. As a result, industry is different and the current industrial profile has important implications for the United States and its global competitiveness.

A Changed Industrial Profile

In the late 1950s, graduating from college with a degree in physics guaranteed anyone a job. Indeed, upon graduation, I received immediate job offers from companies with whom I had not even interviewed. U.S. basic research organizations were the envy of the world.

Huge industrial laboratories at GE, RCA, and Bell meant equal-

> **If your research dollars are going into basic science and your competitor is putting all of their dollars into product development, you are at a distinct disadvantage.**

ly huge basic science research opportunities, and the sheer level of effort involved created confidence that the resources would be available to pursue innovative ideas. By the mid-1970s, Bell labs alone employed some 25,000 scientists and engineers. But all of that has changed. Many of the major industrial labs are gone or much reduced, and industrial R&D is now a more dispersed corporate function. Why has the current massive downsizing of industrial R&D capacity occurred?

First, costs escalated, and companies could not afford the concentrated research power housed in many of the earlier labs. Second, markets themselves became global, and U.S. companies had to compete with overseas companies that did not have basic R&D activities, and did not have to carry that cost into their product price, nor maintain a focus on the long-term at the cost of short-term product development. If your research dollars are going into basic science and your competitor is putting all of their dollars into product development, you are at a distinct disadvantage.

Finally, technology today moves at the speed of light. The most important publishing outlet is no longer Applied Physics Letters. It is e-mail. Large corporate research entities had a difficult time being agile enough to capitalize on their own basic research with sufficient speed.

The Importance of Industrial R&D: The Case of Semiconductors

Who cares? Why does it matter that basic research in U.S. companies is declining? Anyone who bought GE stock in 1965 when the Schenectedy Lab was changed certainly has nothing to complain about. Shareholders have done quite well. But evidence is overwhelming that industrial R&D capacity is critical to continued economic advance. The semiconductor industry offers a case in point.

> **The information age has come about in no insignificant part because productivity in the semiconductor industry has grown at 25% a year for the last 30 years.**

The information age has come about in no insignificant part because productivity in the semiconductor industry has grown at 25% a year for the last 30 years. Every year you can buy a function for 25% less than the year before. Such productivity growth has driven not only expansion in the personal computer industry, but in communications and every other sector in which electronics plays a role.

In short, the $200 billion semiconductor industry has fueled a trillion dollar electronics industry. Currently, the semiconductor industry is growing at 15% a year, which means it will double every five years. If this trend were to continue consistently, semiconductors would be a $2 trillion industry by 2010. These growth rates may or may not be achieved, but growth at some level is sure to characterize the future of the industry. In the United States, it is important to realize that this industry was driven nearly entirely from industrial laboratories. The discoveries, innovations, and manufacturing technologies were produced by the likes of Bell Labs, Fairchild Labs, IBM, Texas Instruments, Phillips, and other companies, some of which do not even exist today.

Yet by the mid-1980s, the U.S. semiconductor industry was on its way to becoming an endangered species. The industry established several cooperative efforts in research, and pursued partnerships with universities. But progress was difficult. In 1987, the

> **With few exceptions, the semiconductor industry in the United States is a product of industrial laboratories.**

industry formed SEMATECH with federal matching funds to reestablish the nation's semiconductor dominance. Today, the industry has won back market share, and there is no federal money in SEMATECH.

So given the growth of the industry and its ability to deal with hard times and strengthen its global position, what conclusions can be drawn from the semiconductor case about the state of R&D in the United States?

First, and most importantly, we as a country do not need to spend more money on R&D. Even though Japan plans to double government-funded R&D over the next five years, and even though they will then outspend us on a dollar basis (albeit not on a per capita or GDP basis), we do not need to pour more money into R&D. We should benefit from the Japanese investment wherever we can. But the $100 billion we spend on industrial research and the $70 billion in government-funded research is probably enough.

What is needed is a plan to redirect that investment. Moving funds out of defense research provides an opportunity to expand support of basic science in non-defense areas. But any attempt to redirect R&D funding will require a road map to priorities. Moreover, the opportunity is before us to develop a priorities road map for basic sciences: physics, chemistry, mathematics, and biology. It is not acceptable for scientists and engineers to argue that science can not provide such a prioritization, that the progress of science is and must be serendipity. If scientists and engineers do not determine and articulate priorities in basic science in the process of redirecting R&D investments, then senators, representatives, cabinet secretaries, and other agencies will.

A unique opportunity is before us to cooperate across disciplines and industrial groups on precompetitive science and technology that will provide innovation into the next millennium and secure a firm competitive position in the global market. The precedents are impressive. The textile

> **As a country, we do not need to spend more money on R&D. We need to redirect the resources we are already spending.**

and apparel industry has created and pursued such collaborative priorities; the automotive industry has; the semiconductor industry has. All are on growth paths. It is time to take this path in other areas of science and industry, to reduce the costs of technology development, and hence continue industry growth.

A Cooperative Illustration of the Case for Optimism

One final illustration makes the point. One of the factors behind the growth in productivity of the semiconductor industry is the progression toward larger wafers. A wafer is a circular disc of silicon on which are located several hundred microchips. In the late 1950s wafers were one inch in diameter. Soon they will be twelve inches in diameter. Obviously, you can make many more chips on a twelve inch wafer than a one inch wafer. In fact, today's microprocessors and memory chips would not be possible on one inch wafers. But an investment is required. One estimate is that conversion from 200 millimeter to 30 millimeter (from 8 inch to 12 inch) wafers will cost about $10 billion, not including factories or equipment. The new

> **When great innovation is possible but comes with a high price tag, scientists and engineers as a group must set priorities. If they do not, then senators, representatives, cabinet secretaries, or someone else will.**

production factories themselves will cost about $2 billion each. No one company controls that level of investment resources. Such huge investments will take cooperation, both in terms of developing the technical capacity and in terms of financing the ultimate operation.

For the first time, such cooperation is a reality. Thirteen international companies—three from Korea, three from Europe, one from Taiwan, and six from the United States—are collaborating over the next two years to develop the equipment necessary to manufacture microchips on the larger wafers. If such collaboration is possible in an industry as competitive as the semiconductor industry, and if we can bring together cultures as diverse as Korea, Taiwan, Europe and America, R&D cooperation can be a reality in many other settings. ∎

V

Negotiating Global Relationships

Technology, as both the pathway and the content of global trade, quickens the pace of economic exchange, broadens the industrial base of goods and services, and blurs previously sharp economic and geographic boundaries. Technology knits the globe together. Some 13.5 million people have access to the Internet. A new site is registered on the Internet at the rate of one every two minutes of the business day. Between 1980 and 1992, the world's civil aviation system nearly doubled its total international kilometers flown, and now carries over 300 million passengers and 142 million tons of cargo on international flights each year. The electronic system for interbank payments processes $2 trillion in bank-to-bank transactions each day. But technology can also be the source of debate and dispute that pulls at the seams of global relationships. Trade disputes, human rights issues, privacy and secrecy interests—all increasingly pivot on technology and innovation. The essays that follow discuss these issues around the world.

R&D Subsidies and Countervailing Measures

Based on a paper delivered to the Center for Science, Trade, and Technology Policy of George Mason University in 1995.

ROBERT K. MORRIS
Senior Policy Director, National Association of Manufacturers

K now thyself," the first inscription at Delphi should also be the first commandment for those who attempt to advance U.S. interests through trade negotiations. In fact, for negotiators the commandment is even more difficult: "know thy country," an entity over which neither the negotiator nor the administration he or she serves can have more than the smallest illusion of control. Consequently, the knowledge required to represent U.S. national interests in international trade negotiations is more a matter of analysis and luck than of stewardship or vision.

I believe that the Uruguay Round of Multilateral Trade Negotiations, concluded in December of 1993 under the auspices of the GATT, was a good agreement for the United States.

Furthermore, the Uruguay Round Agreement on Subsidies and Countervailing Measures, which was a critical element of the overall package, represents an improvement in the international trading rules in these areas and a net gain for the United States.

> The maxim "nothing is agreed until everything is agreed" was particularly apt for the subsidies negotiations, which contained high priority objectives and significant risk for both the United States and its trading partners.

My focus here is on the treatment of research and development under the subsidies agreement. Accordingly, we will deal more with the the costs and concerns associated with the agreement than the expected benefits. Agreements are, almost by definition, compromise documents. Thus the fact that there are disappointments for the United States in this or any agreement does not necessarily imply criticism of the U.S. negotiating effort at Uruguay. U.S. negotiators were outstanding from start to finish. Yet if they can be faulted for anything, that fault lies in not knowing the United States well enough, rather than in any weakness or lack of resolve *vis à vis* our WTO trading partners.

The Framework of the Subsidies Negotiations

It is worth briefly recapitulating the subsidies negotiations and reviewing ways in which those negotiations changed during the Uruguay Round years.

The Uruguay Round was launched in Punta de Este, Uruguay, in September of 1986. From the outset, the plan was to conclude within four years, with a mid-term review in December, 1988. In the area of subsidies and countervailing measures, the negotiators started with a failed document—the

1979 Subsidies Code. From the outset it was understood that the success or failure of the subsidies negotiations could affect the outcome of the entire effort.

One of the most frequently heard phrases during the course of the Uruguay Round was: "nothing is agreed until every-thing is agreed." This maxim was particularly apt for the subsidies negotiations, which contained high priority objectives and significant risk for both the United States and its trading partners, especially countries in the European Community (EC). The challenge for both sides was to achieve a more pragmatic and common view of subsidies and its relationship to trade, while protecting existing polices and practices. Initially, the United States and its trading partners had little in common on this issue, and there was very little progress in the subsidies negotiations in the first year and a half of the Uruguay Round. By the summer of 1988, the battle lines between the United States and the rest of the world were clear.

Philosophically, the U.S. view was that subsidies for the production of traded goods were a blight on the trading system and should be reduced or eliminated. Not counting agricultural subsidies, which were dealt with in a different negotiating group, the assumption behind the U.S. view was that the United States did not use subsidies as a tool of domestic policy; only other countries (the bad guys) did that.

Unfortunately, at the early stages of negotiation, the issue was not subsidies but countervailing duties. On this issue, the United States was on the defensive because it is the only country to use

> Philosophically, the U.S. view was that subsidies for the production of traded goods were a blight on the trading system and should be reduced or eliminated.

countervailing duties against subsidized imports. This fact was important in 1988 because it explained the United States's defense of its GATT-sanctioned countervailing duty regime against a world eager to weaken it.

Later on, however, U.S. countervailing measures had a different significance, which begged the following question: If there is no history of other countries using countervailing duties against imports from the United States or other countries, does the United States have much to fear on this score? In 1988, however, this question did not emerge and the challenge for U.S. Trade Representative Clayton Yeutter and his team was to alter the negotiating dynamic so that more of the focus was on subsidies and less on countervailing duties.

Yeutter's impressive efforts were successful and were reflected in the subsidy component of the understanding reached at the Montreal Mid-Term Review of December 1988. At Montreal, the Gatt Contracting Parties agreed to continue their discussions of the subsidies issues on the basis of a classification system for subsidies generally referred to as the traffic light model. All subsidies were classified into one of the following three categories: (i) prohibited or red light subsidies; (ii) allowable but actionable subsidies, referred to as the amber of yellow light category; and (iii) non-actionable subsidies, the so-called green light category.

Challenges to U.S. countervailing duty laws continued throughout the negotiations, but from the Montreal Mid-Term Review on, much of the subsidies negotiations were devoted to giving practical effect to the traffic light categories. The United States wanted to expand the list of prohibited—red light—subsidies. In this it was successful but only moderately so. Others, notably the European Community, were intent upon creating a

GATT regime that explicitly allowed for protected certain classes of—green light—subsidies. Arguably, the Europeans were more successful.

Research and Development

> Distinguishing between research and development is no easy matter. What may seem applied research or development to a policy maker might be very basic indeed from the point of view of the person or group conducting the research.

Research and development (R&D) were among the subsidized activities for which the EC sought special protection. The U.S. effort to fashion a response centered around three sets of questions, each with its own period of special prominence. The first set of questions were definitional and came to the fore relatively quickly. The second set, which came to dominate negotiations in the latter stages, was sectoral and concerned the effects any agreement might have on individual industries, such as aircraft. The third set of questions concerned how U.S. assistance programs would be affected by a new international regime.

Questions of Definition

The definitional questions were easy to state but hard to answer. What is research? What is applied research? What is basic research? And what is development? In an October, 1990 report on industrial subsidies negotiations, the ACTPN endorsed the concept of a safe harbor for subsidies for basic research. Subsequent advice went a step further, suggesting that there should be a single category for basic and experimental research and that no limitations should be set on the percentage of government funding for a particular nonactionable research project. The ACTPN took a dramatically different approach toward subsidies for developments, suggesting in the October 1990 report that "certain of these [development] subsidies should be prohibited."

Even as members of this group were recommending dramatically different treatment for basic R&D, they were acutely aware that distinguishing between research and development is no easy matter. What may seem applied research or development to a policy maker might be very basic indeed from the point of view of the person or group conducting the research. ACTPN also recognized that these perceptions could vary dramatically from one sector to another. With these concerns in mind, the 1990 ACTPN report recommended an immediate study:

> Given the complex definitional issues involved in research and development, the ACTPN recommends that the White House Office of Science and Technology Policy and the Commerce Department's Office of Technology Policy jointly review the implications of a GATT prohibition for certain development subsidies. These agencies, in conjunction with the interested parties in the private sector, should seek to produce a workable description of those development-related subsidies which, in their view, should be prohibited because of their trade distorting effects.

This writer is unaware of any such study having been produced, but key points had been registered.

As for the actual results of the wrangling over the R&D language of the Subsidies Agreement, the language can be quoted and described but its likely effects are almost impossible to judge. Empirically, there is nothing to go on. As of the summer of 1995, no country had notified the WTO of its intention to offer a subsidy under the relevant provisions of the Subsidies Agreement (Article 8.2 (a)), nor had any invoked the protections of this article in any countervailing duty cases. This is in spite of the fact that the language of the WTO Subsidies Agreement contains the offer of non-actionable status for subsidies that provide "up to 75 percent of the costs of industrial research or 50 percent of the costs of pre-competitive development activity," with the proviso that these funds be limited exclusively to certain specified activities.

As for ACTPN's notion that there is no need for a cap on allowable research, its only expression lies in one of the several footnotes to the agreement. In any case, it is unlikely that the monetary caps will be the most energetically contested aspect of the R&D component of the WTO Subsidies Agreement. That honor is likely to go to the definition of "pre-competitive development activity." This definition, contained in footnote 29 of the Subsidies Agreement, is worth quoting from:

> The term "pre-competitive development activity" means the translation of industrial research findings into a plan, blueprint or design for new, modified or improved products, processes or services, whether intended for sale or use, including the creation of a first prototype which would not be capable of commercial use.

Exactly what sort of prototype might be allowed under this formula is something we can only speculate about at this stage.

The Sectoral Irony

U.S. producers of large civil aircraft were unlikely to view the Agreement's language on pre-competitive development activity as an adequate safeguard. Indeed, having watched the European Community pour $26 billion into the development of Airbus, the affected U.S. companies were determined not to create new loopholes for aircraft subsidies in the new subsidies agreement.

For that reason alone, much of the U.S. business discussion of the R&D provision of the Subsidies Agreement, as well as the discussion of the overall agreement, was heavily influenced by the concerns of the U.S. aircraft producers. Doubtless the aircraft industry's concerns affected the R&D elements of this agreement, and yet the industry itself is not greatly affected by them. The second footnote applying to the R&D language of Article 8.2 (a), declares that the R&D provisions do not apply to civil aircraft. The overall WTO Subsidies Agreement does, of course, apply to aircraft.

Protection of U.S. Programs

The protection of U.S. industrial subsidies programs was initially not an issue of much concern to U.S. negotiators. Yet in the summer and fall of 1992, the issue came alive. It was the subject of meetings organized by the Industrial Research Council, and it surfaced in the fall of 1992 in meetings of the ACTPN Task Force on Industrial Subsidies.

> **Representatives of the technology community within the U.S. government were concerned that an improperly drawn agreement might discourage companies from cooperating with the government**

One consequence of this concern was that government experts who previously had not been involved in these negotiations began to see the subsidies negotiations not only as important but as a threat to existing programs and practices. Officials from the Commerce Department, the Department of Energy, and the White House Office of Science and Technology Policy, all began to make the point that the United States provides a fair number of subsidies that might be affected by the new Subsidies Agreements.

Briefly stated, these representatives of the technology community within the U.S. government were concerned that an improperly drawn agreement might discourage companies from cooperating with the government on new technology projects. They also feared that the process of complying with the requirements of the proposed WTO Subsidies Agreement could provide competitor nations with a snapshot of U.S. goals, policies, and practices.

These concerns did not fall on deaf ears. In defending the Agreement before the Senate Finance Committee in March 1994, Deputy U.S. Trade Representative Ambassador Rufus Yerxa cited the Clinton Administration's effectiveness in protecting U.S. programs:

> *I would also like to make the point that failure to include the R&D green category would have placed at risk a number of very important [U.S.] technology initiatives. . . .*
>
> *I believe that in 2 years' time, were we not to have the protection of this green category, the administration would be back before this committee trying to explain why WTO panel rulings had found our cooperative research and development agreements, our advanced technology program, our NIH biomedical research and commercialization program, the SEMATECH program, the clean car program and many others to be inconsistent with our WTO obligations.*

Ambassador Yerxa cannot be faulted for failing to predict the Congressional election of 1994 or its consequences for R&D funding (U.S. subsidies). Indeed, in March 1994 the trend seemed to be going the other way. The promise of the Clinton Administration as well as the history of both the Bush and Reagan administrations suggested more, not less, cooperation between business and government in technology in the years ahead. But the subsequent facts speak for themselves, and the current trend in R&D cooperation is down, not up.

With the exception of funding for the National Institutes of Health, each of the programs referred to in the Yerxa testimony above is either on its way out or under serious attack. To cite two examples: the corporate members of SEMATECH have announced that they will no longer be receiving government assistance, and the House of Representatives voted in early 1995 to cut the Energy Department's Cooperative Research and Development Assistant (CRADA) program by

> **It is hard to resist the conclusion that the original objective of controlling foreign subsidies had more validity for the United States than the later focus on providing user-friendly international rules for government assistance.**

more than 90 percent. At this juncture, it seems fair to say that if the U.S. paid its trading partners anything to protect U.S. R&D programs, it paid too much.

Conclusion

As a general proposition, the larger the necessary number of parties to a law or agreement, the longer it takes to make it a reality. This general principle applies with even greater force to an agreement that contains specific commitments from the various parties. No one should be surprised that it took the Uruguay Round members many years to produce the current Subsidies Agreement.

For the most part, the subsidies that were the subject of this agreement are *domestic subsidies*—i.e., subsidies that are generated and controlled by the political systems in each of the WTO member countries. Domestic politics, of course, move with lightening quickness, whereas international dialogue is more phlegmatic.

From today's vantage point, it is hard to resist the conclusion that the earlier (i.e., 1986–1990) U.S. negotiating mode was the more accurate one. The original objective of controlling foreign subsidies had more validity for the United States than the later focus on providing user-friendly international rules for government's assistance to U.S. industry. That said, insights brought to bear by the U.S. technology community did serve to bridge the philosophical chasm that divided the United States from the rest of the world on the question of subsidies.

The jury is still out with respect to the operation of the current subsidies code. Fearful that, by means of the Subsidies Agreement, the United States might unwittingly be embracing a commitment to industrial policy and heavy subsidization of industry, U.S. negotiators insisted upon a thorough review of the new subsidies regime within the first five years. If we do not like what we see then, we can scuttle the green box, non-actionable section of the Subsidies Agreement (Article 8.2) as well as the clearly valuable, serious prejudice language of Article 6. This provision gives U.S. exporters important new tools for limiting subsidized competition in U.S. export markets. These decisions will be made by the year 2000. The problem is that, at the current rate, there may be very little experience to draw upon in making those judgements. If so, the United States and its trading partners would be wise simply to extend the deadline. It was too short to begin with. ■

Developments in Global Trade Negotiations: A Comprehensive Subsidy Code

Based on a paper delivered to the Center for Science, Trade, and Technology Policy of George Mason University in 1995.

MARK BOHANNON

Chief Counsel for Technology and Counsel to the Undersecretary for Technology, U.S. Department of Commerce

T he Uruguay Round Agreement is the most comprehensive trade agreement ever. This historic agreement cuts global tariffs, protects intellectual property, disciplines agricultural and industrial subsidies, and creates an effective dispute settlement mechanism.

One area that has received attention since the final round came to a close in December, 1993, is the Agreement on Subsidies and Countervailing Measures, and particularly one aspect of the code, the "green light" for certain research and development (R&D) and development activities.

In looking at this issue, four key points emerge. First, the Uruguay Round Subsidies Agreement established the strictest subsidy discipline ever on all members of the new World Trading Organization (WTO). There is widespread support for what was achieved in the Subsidies Agreement.

Second, the United States has provided, and continues to provide, more support to industrial R&D than any other country. U.S. investment in technology, reflected in longstanding bipartisan support for R&D programs, has contributed significantly to continued economic growth and the creation of jobs.

> **The United States has provided, and continues to provide, more support to industrial R&D than any other country**

Third, the language related to R&D in the Subsidies Code was drafted with U.S. programs in mind. The draft language found in the earlier Dunkel Text tied U.S. hands when it came to working with industry, while leaving other countries' programs safe from action by U.S. countervailing duty laws.

Fourth, there is little or no danger that this provision will become a loophole under which countries will be able to provide production or marketing subsidies. To ensure that these provisions are fully and fairly carried by all signatories to the GATT, the effective monitoring of all the parties is key to implementing these provisions.

An Historic Agreement

It is worth reviewing the achievements of the GATT negotiations. Both workers and industries benefit from the GATT Agreement negotiated at the end of 1993. Significant new employment opportunities and additional high-paying jobs in export industries will emerge. U.S. business will directly benefit from emerging opportunities to export more products in the agricultural, manufacturing, and service sectors. Specifically, the agreement will:

- cut foreign tariffs on manufactured products by over one-third, the largest reduction in history;
- protect the intellectual property rights of U.S. pharmaceutical and software industries from theft in the global marketplace;
- ensure open foreign markets for U.S. exporters of services such as accounting, advertising, computer services, tourism, engineering, and construction;
- greatly expand export opportunities for U.S. agricultural products by limiting the ability of foreign governments to block exports through non-tariff barriers, quotas, subsidies, and a variety of other domestic policies and regulations;
- protect the right of the United States to provide relief from unfairly traded imports;
- assure that developing countries live by the same trade rules as developed countries; and
- create an effective set of rules for the prompt settlement of disputes.

Strict Subsidies Discipline

The subsidies agreement establishes a three-class framework for the categorization of subsidies and subsidy remedies:

- prohibited subsidies;
- actionable subsidies, which are subject to dispute settlement under the WTO in Geneva and countervailable unilaterally under domestic laws if they cause adverse trade effects; and
- non-actionable and non-countervailable subsidies (the so-

called "green light" category), structured according to criteria intended to limit their potential for distortion.

The strict new disciplines and effective new dispute settlement system includes a number of major changes that will benefit the United States:

- The agreement will apply to all 117 members of the World Trade Organization, a vast improvement on the Tokyo Round Subsidies Code, which had only 27 signatories.
- For the first time in the history of GATT, a subsidy is defined and the conditions which must exist in order for a subsidy to be actionable are set out.
- The agreement extends and clarifies the 1979 Subsidies Code's list of prohibited practices to include *de facto* as well as *de jure* export subsidies, and subsidies contingent upon the use of local content.
- The agreement also specifies how to prove "serious prejudice" (adverse effects) to a country's trade interests and creates an obligation for the subsidizing country to withdraw the subsidy or remove the adverse effects when they are identified. The absence of such a provision in the 1979 Subsidies Code was one of its greatest deficiencies.
- The agreement introduces a presumption of serious prejudice in situations where the total *ad valorem* subsidization of a product exceeds 5 percent, or when subsidies are provided for debt forgiveness.

Integrating Technology and Trade

The language that was negotiated in the final round reflects both strong trade policy and competitive technology policy. These provisions will enable the United States to fight unfair subsidies that distort free trade, while at the same time protect U.S. firms that participate in technology programs here at home.

U.S. negotiators started with the universal view that the draft Dunkel Text presented a number of concerns for the private sector and for our commitment to public–private partnerships on technology, while achieving none of our trade goals.

R&D Infrastructure

The Dunkel Text undercut one of the primary advantages the United States has over its competitors: R&D infrastructure. The United States has been, and continues to be, the greatest supporter of industrial research in the world. In terms of total government R&D expenditures, the U.S. invests four-and-one-half times the amount of our closest competitors—Japan and Germany. Even when defense-related R&D is excluded, the U.S. spent $28.9 billion on civilian R&D in 1991. Germany, the next largest country, spent 55 percent less than the U.S. on civilian R&D.

> **The Clinton Administration has reinvigorated the public–private partnership as a key means of enabling technologies that are key to economic growth.**

The above figures are the latest evidence of a long-term, bipartisan commitment to technology investment to promote economic growth. The tangible examples of these investments include programs like the Advanced Technology Program at the Department of Commerce, as well as the dual-purpose initiatives embodied in the Technology Reinvestment Project at the Department of Defense. They also include the world-class biomedical research of the National Institutes of Health; Defense Department investments in flat panel displays and multi-chip modules; and an increased focus on civilian technology by the national laboratories. The U.S. commitment to technology investment through public–private partnerships is also reflected in the more than 2,000 Cooperative Research and Development Agreements that have revolutionized industry–government collaboration.

The Clinton Administration has reinvigorated the public–private partnership as a key means of achieving technology investments. The administration supports a policy that requires projects to be cost-shared (often 50 percent from industry and 50 percent from government), and the selection process merit-based. These initiatives reflect the proper role of government in working with industry to sustain the high-risk, enabling technologies that are key to economic growth.

Under the draft Dunkel Text, the more transparent U.S. technology programs would have been open to foreign challenge. It would have impeded what every administration has recognized: investment in research and development is a desirable, effective, and long-term investment in our future.

Basic and Applied Research

A second problem posed by the draft Dunkel Text was that it relied on definitions of "basic" and "applied" research that did not fit the model of U.S. technology programs. This ambiguity was compounded by the fact that thresholds of non-actionable government investment envisioned in the Dunkel Text were out of line with the fact that programs should be equally cost-shared.

Private Sector Concerns

The private sector was deeply distressed with the Dunkel Text's provisions related to "notification." In order to gain limited protection under the Dunkel Text, highly detailed notifications of programs would have had to be made to the GATT Subsidies Committee, possibly requiring the government to share extensive and competitively valuable information about activities of U.S. firms. Instead of seeing hope and protection in these notification requirements, the private sector saw greater regulation, more paperwork, threats to sensitive information, and less incentive to work with government in this important area.

The portion of the Uruguay Round Agreement that addresses R&D investment is a major improvement over the Dunkel Text, from both a trade and technology perspective. U.S. investment in "fundamental research" is fully protected.

U.S. negotiators have ensured that government involvement in "industrial research," a mainstay of our public–private partnerships, continues without threat. The government may be involved, either directly with funds, personnel, or in-kind resources in critical investigations aimed at the discovery of new knowledge, with the objective that such knowledge may later be useful in developing or improving new products, processes, or services. These kinds of partnerships are industry-focused and pre-competitive, and have the potential to provide benefits across a number of companies and industries.

Consistent with such bipartisan, merit-based, cost-shared technology programs, government may partner up to 50 percent of a project that focuses on "pre-competitive development activity."

The Uruguay Round achievement also addressed the sensitive issue of "notification." The agreement maintains the ability to provide protection through special notification, but it does not mandate that such notification occur in order to protect an investment from trade measures under the R&D criteria. Instead, if there is ever a challenge, countries can at the time show how any support provided is consistent with the R&D provisions. The final Uruguay Round text also clarifies that the notification requirements will not force the U.S. to release any proprietary or confidential information to the GATT Subsidies Committee.

Conclusion

Had the United States not sought changes to the "green light" rules governing R&D, the result would *not* have prevented or discouraged foreign governments in their support for industrial research and development. Instead, our European trading partners would have enjoyed the protection of the Dunkel Text's green light rules, which were patterned after the European Community's own internal rules, while U.S. technology programs would not have enjoyed such protection.

> The Uruguay Round represents an integration of trade and technology policy, an important facet of which is a continued commitment to fight unfair subsidies used by other countries.

The completion of the Uruguay Round represents the latest step in a long-term bipartisan effort to improve the world trading rules and enhance U.S. competitiveness. It represents an integration of trade and technology policy, an important facet of which is a continued commitment to fight unfair subsidies used by other countries. ■

Trade in Financial Services and Advanced Technology: Clear Issues but Elusive Policy

Financial Services and Open Trading Markets

Summary of remarks delivered at the New York Academy of Sciences.

F. WILLIAM HAWLEY

Director, International Government Relations, Citicorp/Citibank

he seeds of involvement in the Global Agreement on Tariffs and Trade (GATT) by financial services industries were sown many years ago.

A History of Protectionism

The financial services industry, which includes banking, insurance, securities, and diversified financial services companies, began to observe the emergence of various types of sophisticated barriers to outside trade in financial services. The effect was to protect the small community of local financial institutions by placing restrictions on the activities of foreign financial companies.

> **Extending trade negotiations to include such areas as finance and telecommunications is critical to keeping the GATT itself relevant to the composition of today's global economy.**

This pattern was problematic for the U.S. financial industry. But equally it was inimical to the economic growth of countries who practiced protectionism. In effect, the growing financial protectionism was denying these economies access to the competitive capital sources that they needed to continue to develop.

Hence, the financial industry attempted to persuade GATT negotiators to include financial services within the framework of GATT negotiations. Although services had been left out of the structure of GATT at its inception in 1947, today the services sector represents a majority of jobs and GNP in the United States, and is increasingly important throughout the world.

Why GATT as the Solution?

With the onset of the Uruguay Round of GATT negotiations, four key reasons for including financial services appeared.

For many years, the U.S. Treasury Department had negotiated financial disputes through bilateral agreements. This approach was increasingly ineffective. The United States does not have constant bilateral trade consultation with most countries of the world. So as the protectionism trend spread, the bilateral mechanism could not be expected to keep pace.

A second problem was that focusing exclusively on financial services in existing bilateral talks provided too narrow a base for negotiations. By its very breadth, the GATT held the potential to provide an enormous range of "carrots and sticks" across the complex of economic sectors in the negotiations. Hence, accommodation on service sector issues could be matched with negotiations in other sectors or on other issues so that the entire package of compromises, taken as a whole, would be more broadly acceptable.

Third, the result of a successful broader approach to negotiations would be more than the sum of its parts. The result would not simply be a set of rules to stop proliferation of trade restrictions, it would also be an impetus toward real market liberalization around the world. In turn, this would be advantageous for both U.S. industry and for overseas economies of all stages of economic development.

Finally, including services industries in the GATT would keep GATT itself relevant to the composition of today's global economy. With the increasing importance of trade in financial services, it is essential to include services if the GATT is to play a trade expansion role into the 21st century.

Why Financial Services Negotiations Failed

Despite the general success of the Uruguay Round of the GATT, the negotiations did not achieve their agenda in the area of financial services. The failure can probably be traced to a series of early mistakes.

Initial negotiating emphasis was placed on creating a framework agreement for financial services with airtight definitions. The agreement was to be based on the concept of "national treatment"—i.e. equal treatment of both foreign and domestic firms. With such strict language in place, those who signed the ultimate agreement would be bound to obey its principles and provisions.

But it quickly became clear that virtually no one would even talk about such an agreement. The fallback approach was to develop a non-binding set of principles. In this mode, each signatory merely created a separate annex that specified those areas in which they were willing to specifically apply the general principles. In effect, the approach was transformed from one in which there are specified exceptions to the agreement (a "negative list," in trade parlance), to one in which nothing is binding beyond a list of exceptions (a "positive list").

Much of the hesitancy to engage in specifics in financial services was a fear that such trade negotiations would interfere with the traditional rights and responsibilities of national banking regulators to do their jobs, and national financial institutions to affect overall fiscal and monetary policy.

As a result, the framework agreement resulted in no country being committed to any specific list of binding actions.

Hard Lessons from Hard Experience

The Uruguay Round experience illustrated the complexity of conducting financial services negotiations. A number of key lessons emerged.

First, it is critical that financial services negotiations be placed within a larger context of negotiations from the outset. The broad sectoral array of wide-scale negotiations is needed to force compromises on financial services issues because it is such a sensitive area of national interest.

Also, flexibility and creativity is needed to accommodate the various levels of development of financial sectors across a range of countries. In the North American Free Trade Agreement (NAFTA), for example, such a flexible approach was successfully taken. The opening of financial markets implicitly recognized that Mexico was just beginning to privatize its financial sector, and hence the trade liberalization was phased in and placed under the clear control of the Mexican government. This type of structured flexibility may prove critical in the next phase of financial services negotiations globally.

The Elusive Scope of the Financial Services Industry

STEPHEN A. HERZENBERG

Office of Technology Assessment, United States Congress

T he service sector accounts for over 80 percent of the U.S. economy and has long deserved more attention than it has received in both economic and trade circles. The Office of Technology Assessment has been looking at a variety of issues in this area.

With specific reference to financial services, a number of difficult questions must be answered if the full role of this industry is to be understood and its importance guarded.

Problems in Productivity Measurement

If productivity growth in services industries is hard to measure in general, it is particularly difficult in financial services. The proliferation of new products and improvements in service quality made it extremely hard to track output in a constant, meaningful way. It is fair to say that we are certainly underestimating productivity growth in most of the financial services industry.

Flattening the Service Job Hierarchy

A second difficulty is the broadening of jobs at the bottom of the employment hierarchy, a trend that is common to the financial services and telecommunications industries. Line workers are tied into sophisticated data bases and now can provide a wide range of services to a large number of customers. The merging of job responsibilities within the employment hierarchy affects the organization of work. It may also affect employment levels. These changes in employment organization can increase industrial competitiveness and

> **Services negotiations in trade agreements will need to confront issues of wage and labor standards or the polarization that characterized the NAFTA debate will be replicated repeatedly.**

enhance the importance of service industries in global trade negotiations.

The International Mobility of Service Sector Jobs

The international mobility of manufacturing jobs has long been a subject of global economic debate. Such mobility may also become an increasingly important issue in service industries. With innovations such as electronic data interchanges and 24-hour centralized loan application centers, questions of job mobility in the financial services industry arise. How mobile are these functions now within the United States?

And how mobile are they internationally? Indeed, the export of low-wage clerical jobs in service industries may prove to be one of the motivations for less developed countries to retain liberal trade policies in service sectors such as finance. While international out-sourcing of clerical jobs is more of a trickle than a flood at the present point in time, the issue may grow as technology innovation—such as the proliferation of optical character readers—changes the nature (and perhaps the international mobility) of many jobs.

Learning from NAFTA

Our growing understanding of the trends in financial services industries suggests that one of the elements of trade negotiations in services will encompass labor rights. Under the NAFTA negotiations, of course, the issue of labor rights became very controversial. The controversy spills over into issues both of economics and of social standards. Hence, it can be very explosive and greatly contribute to national tendencies toward protectionism.

It is clear that productivity and quality in the service work force in the United States is rising. It is also clear that there is a disparity in the work force that increasingly cleaves along skill and education lines. Less educated workers in this country today suffer much higher rates of unemployment and underemployment than historically has been the case. This same trend, although less pronounced, can be seen in other parts of the world.

In many instances, it is this fear of being left behind that has driven protectionism, whether among French farmers protesting the European Union or U.S. unions confronted with the NAFTA negotiations. If new agreements in trade liberalization are the objective, then financial services industries and technology-based companies would be well advised to recognize the wage and employment concerns that have driven protectionism in manufacturing and agricultural sectors and be prepared to address these concerns in service negotiations.

Telecommunications: The Nervous System for Open Financial Markets

WALTER I. RICKARD
Group Vice President, NYNEX

T he financial services and telecommunications industries are tightly linked. Discussions of financial services policy cannot be usefully undertaken without also discussing telecommunications policy.

The Promise of New Technologies

The world is entering an unprecedented age of integrated communications between people, between enterprises, across organizations, and among all nations. Digital, fiber, radio, and satellite transmission will combine, and accessibility of all types of communications for all types of purposes will pervade global communications.

These advances will permit significant gains in the speed with which financial services are provided. Worldwide financial transactions will take place in less than six seconds. Advances will also affect the location of financial transactions. Consumers will be banking via their television sets, personal computers, or data phones. This is a new information architecture that will be able to seamlessly interconnect everyone anywhere in the world.

Eight Principles for Expanded Trade

For a global architecture to work, however, open market access throughout the world will be essential. In moving technological innovation toward such an open system, eight key principles must form a clear core of the trading system.

First, open market access must be assured. Telecommunications enterprises must have the ability to build, own, operate, and re-sell

> Information technologies and providers represent a new architecture for seamlessly connecting people and institutions. But a global architecture requires global open market access.

technology networks worldwide.

Second, national treatment standards must be applied. Foreign telecommunications companies should be treated just as their domestic counterparts in any country in the world. This principle should be applied equally to the United States market.

Third, safeguards must be established. Telecommunications services must be unbundled, interconnections among services encouraged, and a common set of cost accounting rules applied.

Fourth, standards must be open and integrated. The market place is moving forward much faster than our international standard setting process. This is certain to impede investment and trade.

Fifth, deregulation is essential. If technological innovations are to be maximized in the market place, investment must be encouraged. And investment requires a much more encouraging regulatory environment in many countries.

Sixth, trade regimes should encourage international partnerships. The architecture can and should be seamless, and that means that interconnection arrangements must be open.

Seventh, national telecommunications networks must be open to foreign participation. If the architecture is to be globally seamless and take advantage of new technological innovations, then national networks must be able to merge.

Eighth, public subsidies must be rethought and retargeted. Telecommunications pricing per minute is always higher if it must support public subsidies. In New York State, for example, a program called "lifeline" provides subsidized telephone service to people on public assistance. The subsidy is paid by higher rates on other customers. The permutations and combinations of such programs in nations around the world are myriad. The result, however beneficial in terms of social policy, is a crazy-quilt in the competitive market place. Trade policy which focuses on telecommunications and financial services will need to address this issue.

Reasons for Hope

The Uruguay Round of the GATT provided several reasons for optimism. Service industries are now seen as a key part of global trade negotiations, and, within these considerations, telecommunications will receive considerable attention.

Moreover, precedents in bilateral arrangements are now serving as prototypes of multilateral negotiations. NYNEX, for example, has an arrangement with the telecommunications agent for Bangkok to build, operate, and own a 2 billion line network in Thailand. NYNEX also has the largest cable franchise in the United Kingdom. Knowledge of successful bilateral efforts in telecommunications will provide considerable guidance to multilateral trade negotiators as to the critical elements necessary to liberalize global telecommunications trade. ■

Reforms in Russia: Emerging Trends and S&T Dimensions

Based on remarks delivered at the New York Academy of Sciences on October 28, 1994. In May 1997, Ambassador Pickering became Under Secretary of State for Political Affairs.

THOMAS R. PICKERING

United States Ambassador to the Russian Federation

I n a country with a history as rich and as deep as Russia's, it may seem premature to offer an assessment of sweeping political and economic reforms that, after all, are only three years old. Such skepticism may be particularly acute because it is barely a year since the post-reform political struggle reached its most intense and bloodiest level. In modern-day Russia, however, three years is equivalent to an era. So an assessment is appropriate and the results are positive. The process may not always be smooth and progress may be uneven, but overall, democratization and economic reforms are working in Russia.

Political Progress . . .

Democracy's roots are growing deeply into Russia's political soil. In December 1993, Russia conducted a fair and free election for

> **The process may not always be smooth, and progress may be uneven, but democracy and economic reform are working in Russia.**

a legislative body and adopted a constitution untainted by the ideology and precepts of the Soviet era. The legislature began its second session on October 5, 1994, a significant milestone given the dire predictions in September 1993 that the dissolution of the Supreme Soviet would only prolong internal political confrontation and erode support for reforms. On the contrary, and despite significant political differences, the Russian legislature has shown that it can work constructively with the executive across a range of areas. Equally significantly, the overwhelming majority of average Russians are adjusting to a new political and economic reality. The majority rejects a return to a totalitarian system and accepts that a representative form of government, with all of its problems, holds the key to a better future.

. . . Combined With Economic Reform . . .

Economic reforms are also taking hold. Over the past year, the Russian government has worked responsibly to control inflation, adhere to budgetary targets, and establish a legal framework for commercial transactions. Russia has completed the first phase of the largest privatization plan ever attempted. Despite great difficulties, Russia has also embarked upon a reform of the agricultural sector.

There are, of course, highly publicized setbacks. Many difficulties, including the threat of crime and corruption, remain for foreign investors. But progress is measurable and evident throughout the country. The stock market continues to flourish. Retail trade is vigorous. Private investment capital has begun to flow into the country. The question is no longer whether economic reforms are real, but how to speed up a process that is well under way and making palpable progress.

. . . Reinforce a Firm Commitment of U.S. Support . . .

The United States continues to support the process of democratization and economic reforms in many ways and at a number of levels. Perhaps the most tangible example of such support is financial assistance to Russia. In 1993–1994, U.S. aid to Russia totalled $2.5 billion; in 1994–1995, U.S. aid programs are projected to total $850 million. The financial role of the United States government, however, is necessarily limited in these days of tight budgets and competing priorities. Ultimately, the engine that must drive economic progress in Russia is not government assistance, but private investment. Toward that end, at the Washington summit in September 1994, President Clinton and President Yeltsin signed a framework agreement for economic partnership that will lower barriers to trade in both countries, and provide a foundation for further private investment and commercial transaction.

. . . Together with an Enduring Concern for Human Rights

In recognition of Russia's commitment to the protection of human rights, best exemplified by its constitution's strong bill of rights, President Clinton has declared Russia in full compliance with the Jackson-Vanik requirements of the Trade Act of 1974, which denied Most Favored Nation Status to those countries who limited free emigration by their religious minorities. This will allow trade between the two countries to flow with consistency and confidence.

> **The Russian government will raise the priority of human rights as society exercises its own rights through the political system.**

This does not mean, however, that the U.S. government's concern over human rights issues in Russia has ended. Significant progress has clearly been made. Scores of political prisoners no longer waste away in Russian gulags. Psychiatry is no longer abused as a political tool. Nevertheless, many of the fundamental human rights guaranteed to Russian citizens in the new constitution exist only as future hopes. Prison conditions are abysmal. Police brutality abounds. Rights of refugees, both internal and external, are overlooked. Discrimination, cronyism, and outright corruption permeate all levels of civic and official structures.

The solution to these and other human rights problems will come only as the process of democratization advances. The Russian government will raise the priority of human rights on its political agenda only as society learns more about its own rights and begins to exercise them through the political system. Thus, U.S. support for democratic reform at the grass roots is an important element of our support for human rights in Russia.

The Future of Russian Science

The great achievements and long traditions of Russian science enjoy world renown. Indeed, it was through science that the strongest voices of protest spoke out in defense of democratic principles and individual freedoms. Perhaps the most notable, but certainly not the only, such leader was Dr. Andrei Sakharov.

Paradoxically, now that democratic freedoms are attainable, the future of Russian science is itself in question. There is widespread debate about whether or not scientific institutions will survive the reform process and, if so, in what direction they will ultimately go. There is a clear answer to the first question: Russian scientists will indeed survive the current financial crisis and political reform process. The "brain drain" process among Russian scientists is a process of emigration from government-supported science in Russia to private sector opportunities abroad. As the reform process in Russia deepens, and renewed efforts to support science take hold, young Russian scientists will return with experience, knowledge, and the proper tools to work in their fields.

The future orientation of Russian science will depend, to some extent, on whether the United States and other Western countries provide the opportunities for cooperation that will enable Russian scientists to emerge from years of isolation and secrecy. That support must come from both government and private sources.

U.S. – Russia Scientific Collaboration

During the 1993 Vancouver Summit, President Clinton and President Yeltsin committed the U.S. and Russia to a renewed partnership on science and technology by establishing a U.S.–Russia Commission of Technological Cooperation on energy and space co-chaired by Prime Minister Chernomyrdin and Vice President Gore. The mandate of the Commission has now been extended to include basic sciences and the environment.

Among the first initiatives of the Commission is closer cooperation in space exploration. Under the space agreement, the U.S. will provide Russia with as much as $400 million between 1994 and 1997 in the form of payments for equipment and services for a joint Shuttle/Mir flight program and the early stages of planning for the space station. These and other collaborative ventures will provide more comprehensive data in such areas as environmental monitoring and global climate change research.

In the area of energy, extensive collaboration on fossil and nuclear energy is under way to identify improved technologies, develop environmental remediation technologies, and assist with the commercialization and privatization of Russian facilities. The

> **The financial role of the United States government in Russian economic reform is necessarily limited in these days of tight budgets and competing priorities. The engine that must drive economic progress in Russia is private investment.**

U.S. and Russia are also working jointly to upgrade the safety of nuclear power reactors in Russia and improve operating procedures, regulatory guidelines, and emergency response capabilities. The legal framework established for these efforts will also contribute to closer private collaboration for civilian nuclear power. The Commission of Technological Cooperation also has established an agreement that, for the first time, opens all fields of science and technology to joint cooperation. The agreement, for example, provides for protection of intellectual property rights in cooperative programs. In addition to S&T efforts within the Commission, the State Department last year disbursed $700,000 to Russian scientists for global climate change projects. The Agency for International Development (AID) also has programs for environmental cooperation, notably in the Lake Baikal region.

The U.S. national laboratories have developed collaborative relationships with their Russian counterparts for projects ranging from basic research to the development of new commercial technologies. These efforts include improvements in the safety and security of nuclear materials, a critical issue in the post-Cold War world.

National Security and Defense Conversion

Defense downsizing and military conversion represent areas of real difficulty for Russia, as they do in the U.S. Whereas the U.S. seeks to reduce defense spending from 6 percent to about 4 percent of GDP, in Russia the numbers are much higher, and the consequent political and economic pain is much sharper. The Soviet economy was overwhelmingly militarized, and today large swaths of that military-industrial complex lie idle. A very large portion of the 50 percent decline in Russian industrial output is attributable to defense cutbacks.

In this context, the United States is helping Russia reduce its stockpiles of nuclear, biological, and chemical weapons, as well as associated missile delivery capability. Four Russian defense firms have already been paired with U.S. joint venture partners for conversion projects. Another eighty-three such firms have been identified for assistance, pending availability of program financing, which is contingent on the quality of their proposals. Finally, the U.S. has funded the Independent Defense Enterprise Fund, which received an initial capitalization of $40 million from Congress to help finance commercially viable defense conversion projects. While U.S. assistance is important, the market will ultimately determine which Russian defense firms survive.

New Collaborative S&T Institutions

Further evidence of international collaboration to secure Russian science can be found in two new institutions. The International Science and Technology Center, jointly funded by the U.S., Russia, Japan, and the European Union, has already funded some seventy-six proposals for non-military research by former weapons program scientists. Private sector commitment has also been critical. The greatest contribution has been made by the International Science Foundation, established in late 1992 with start-up funds of over $100 million from George Soros to promote basic science research. Chaired by Dr. James Watson of the Cold Spring Harbor Laboratory, the ISF has made thousands of individual and institutional grants for research and funded the travel of hundreds of

Russian scientists to international conferences. The Russian and American governments are now discussing ways to work together to continue this initiative.

The private commercial sector is also active in creating new initiatives. Various U.S. companies are joining with Russian scientists to establish new centers for R&D that are profitable and mutually beneficial. The opportunities are real, and the investments will pay off.

A Time For Action

Russia is well on its way to creating the conditions necessary for successful economic reform and thriving investment. But we cannot sit back and wait for the perfect time for further support to arrive; the perfect time for action will only come with our involvement. The United States has a proud record of supporting Russia in political partnership, security consultations, and scientific collaboration over the past two and a half years. The coming years will be equally critical, and success will be ensured only by collaboration between Russia and the United States and between the private and public sectors within Russia. ■

China Changing: Can the U.S. Support Human Rights Without Sacrificing Sino-American Relations?

Based on remarks
delivered at the New
York Academy of
Sciences on November
3, 1994. In August
1995, Dr. Oksenberg
moved to Stanford
University, where he is
professor of political
science and senior fel-
low at the Asia Pacific
Research Center.

MICHEL OKSENBERG

President, East-West Center

T
he search for an intersection between steadfast support for human rights abroad and the economic and security interests of the United States is neither a new nor an easy undertaking for U.S. policy makers. In the case of China, however, finding that intersection is of critical and immediate importance. China is, and will remain, a significant, even massive, global force. Indeed, the world faces a range of issues that cannot be successfully addressed without China's active involvement. Fortunately, there is a great deal of compatibility between the pursuit of human rights and the pursuit of many other objectives of the United States with regard to China.

Defining U.S. Interests

It is important, first of all, to be clear about the definition and range of U.S. interests vis-a-vis China. One of the most important is to see the emergence of a China that contributes to the peace and stability of the Asia/Pacific region. To take one example, U.S.-China cooperation is essential to maintaining stability, especially on issues of nuclear proliferation, in Korea, Taiwan, Hong Kong, Southeast Asia, and South Asia.

The successful management of a range of problems confronting humanity in general, and the U.S. in particular, also hinges on Chinese behavior. The environment is an obvious area in which the choices China makes will affect the world's people. Emerging disease patterns also provide sobering examples of problems which defy national boundaries. The progress of communicable diseases, and global prospects for control of diseases such as HIV, will be significantly affected by the control efforts of China. Similarly, the spread of narcotics cannot be slowed, let alone halted, without Chinese cooperation.

In the realm of economics, China has become, and will remain, a global economic and trading leader. China's economic growth has averaged 10 percent a year over the last few years, with some provinces in this vast subcontinent averaging nearly 20 percent a year. This represents the most rapid and extensive economic growth in human history. China is also a critical site for foreign investment and the world's largest market. More than 30 percent of all foreign investment in developing countries in 1993 occurred in China. Much of that investment was U.S. in origin. Moreover, nearly 40 percent of all Chinese exports are destined for U.S. markets. Hence, the United States has an interest in seeing that China's growing economic role occurs in an orderly and positive fashion on the global economic stage.

Human Rights: A Working Definition

There are many within and outside of Asia who would argue that the term "human rights" encompasses a range of human welfare and economic issues. Within this broad context, economic development and social programs become important indicators of human rights assessments.

For the purposes of policy discussion, however, a more useful approach narrows the term to apply to political human rights—the basic yearnings for freedom and security that are not simply Western in origin. Such a definition focuses on fundamental issues such as detention without notification of kin or notification of charges, detention without a time limit of release if no charges are filed, torture, summary execution, and equal access to the law. It is these basic aspects of human rights that are normally linked to U.S.–China economic relations.

Under the broad definition, it must be acknowledged that many conditions in the West, and in America, lead Asians in general to perceive strident human rights advocacy as arrogant. It is important to remember that the human rights restrictions attached to U.S. foreign aid abroad would have prohibited federal aid to the U.S. itself until 1964. Therefore, while the human rights values of the U.S. are noble and worthy, they should be put forward with a certain humility.

Human Rights in China: An Emerging Opportunity for Change

Until recently it has been very difficult to broach basic human rights issues in China with any success. But change is afoot. Indeed, the time has come to pursue an enlightened policy of engagement with China in the sphere of values and institution building. As in many other countries, the scientific community in China would be among the most receptive to such an approach.

> **The time has come to pursue an enlightened policy of engagement with China in the sphere of civic values and institution building. The scientific community in China would be among the most receptive to such an effort.**

The reasons for the emergence of an opportunity for renewed action are many. The logical consequence of China's opening to the outside world has been a renewed recognition of the importance of basic political rights within ever larger portions of the population. While "freedom" would be too strong a term to describe the current state of social organization, there is, nevertheless, a growing open expression of private opinion that drives in the direction of political change. This is not surprising. Economic growth of 10 percent a year cannot help but unleash social forces and popular expectations.

The Importance of Institutional Reform

The very real problem is that fundamental institutions in China remain too weak to adapt to these new expectations and to channel change in the direction of stable and humane government policy. Instead, economic growth has led to a general erosion of authority in China, to corruption, nepotism, and a splintering of political control. In this context, both human rights and economic growth, not to mention any intersection between the two, are at risk. The following examples will serve to illustrate the likely implications of weak institutional development for China's overall future.

The Military

In the realm of national security, civilian institutions for guiding the military are relatively underdeveloped. The formal chain of command runs from the Communist Party's Politburo to the Party's Military Affairs Commission to the chiefs of staff. The military is under Party control, and a major purpose of the military is to keep the Party in power. But the authority of the Party is being eroded by explosive economic growth; hence, the single buffer between the military and civilian rule is weak. There are no alternative institutional mechanisms capable of ensuring civilian control over the instruments of coercion in China. The implications are obvious and probably disastrous.

The Legal System

The legal system in China is another example of an underdeveloped institution. China is basically a country without an extensive legal system or a well-developed, independent judiciary. This state of affairs has deep historical roots: while there is considerable precedent for administrative law within large bureaucracies, the basis of economic transactions was through well-developed interpersonal relations, not by enforceable contract law.

Without an independent, enforceable system of law and adjudication, human rights will never be fully secured. Individual citizens will always be vulnerable to the arbitrary rule of political leaders. Advocacy for human rights, to be ultimately successful, must also become advocacy for the development of legal systems and institutions.

The Environment

China voluntarily adheres to a number of international environmental accords, including the Montreal Accords on the reduction of ozone-depleting substances, and international treaties on the protection of endangered species, ocean dumping, and the preservation of cultural and national heritage. Relative to other developing countries, China is actually in the forefront of global responsiveness to such agreements.

Enforcement, however, is quite another matter. While groups of scientists and policy makers in China are committed to environmental enforcement, institutional mechanisms to implement that commitment are insufficient.

When China enters into a treaty, Beijing provides little or no budgetary support and relies upon the voluntary cooperation of thirty provinces for enforcement. The provinces, on the other hand, are predominantly concerned with economic development. Environmental regulatory enforcement does not rise very far up the provincial priority list.

Weak Institutions = Missed Opportunities

Weaknesses in China's institutional structures limit the opportunities to integrate China more fully into international efforts to resolve global problems. Indeed, given the narrow institutional base in China, successful efforts to assist China in enforcing its global commitments often have perverse effects in other areas, including human rights. A few examples will illustrate the problem.

Beijing has committed itself to the protection and preservation of endangered species. Of particular concern are the Siberian tiger and the rhinoceros. Both have medicinal values in China and Taiwan, especially for the elderly. Therefore, a vibrant market exists for tiger bone and rhinoceros horn. Indeed, tiger bone is part of the Chinese pharmacopoeia. However, under these circumstances, there are virtually no enforcement mechanisms in China with sufficient resources or influence to enforce species protection. The combination of dispersed provincial responsibility for enforcement, insufficient resources to support that responsibility, and a vibrant and profitable market for the product itself combine to limit the opportunity for capitalizing on the global environmental commitment that Beijing has made.

Institutional problems become even more complex when environmental issues cross over into the national security sector. In the Montreal Accords, for example, halons are specified as a major ozone-depleting substance. Halons, however, are also a major substance for fighting electrical fires. In China, the public security ministry is in charge of fire-fighting. Hence, improving Chinese capacity to implement the Montreal Accords entails improving the technological base of the very ministry whose human rights behavior is in question.

Both these examples serve to illustrate the general point that weak institutional development stands in the way of translating the logical social implications of growth and openness into more humane and responsible political structures.

Dividing the Labor of Human Rights: A Role for the U.S. Government

There is, nevertheless, a growing opportunity to work with individual Chinese who are themselves interested in enhancing the effectiveness of institutions. These individuals encompass the Chinese judiciary, parliament, and journalism—areas in which both China and the U.S. have a mutual interest. Well-developed institutions are at the heart of effective governance, which itself is essential to sustaining China's economic growth.

Clearly, China is at a crossroads. The basic question is whether it will emerge as a chaotic, corrupt, disorganized, but still economically vibrant nation with a weak central government, or whether it will develop as an orderly society with a humane government

> **The question is whether China will emerge as a chaotic yet economically vibrant place with a rather weak central government, or an orderly society, humanely governed and responsive to popular will.**

responsive to popular will. In this context, the emphasis of the United States government should be on working cooperatively on the development of a whole range of institutions which underpin a civil society. Unless this task is undertaken soon, a China that resembles the first alternative is more likely than one that resembles the second.

> **Private human rights groups must continue to exert pressure to resolve human rights abuses. The U.S. should initiate complimentary efforts to assist with the creation of the legal, political, and journalistic institutions necessary to extend and ensure such human rights.**

And A Role for Private Advocacy

At the same time, private organizations have an essential role to play in edging China toward developing a more humane and effective government. Past experience has taught that the government-to-government tool becomes a blunt instrument in China when arguing for individual human rights abuses. Government exertion of pressure often results only in deeper distrust, and sets back larger institution-building agendas. Yet, such individual cases exist and must be argued.

Private effort, such as that of the Committee on the Human Rights of Scientists of the New York Academy of Sciences, is essential and often very effective in pursuing individual human rights cases. As China opens, the country's need to be perceived as globally responsible grows. Private organizations, in providing solutions to China on human rights abuses, become a mechanism for making a clear link between China's behavior and its global reputation. ∎

Science and Technology in Hong Kong beyond 1997: Emerging Trends and Issues

Based on remarks delivered to the New York Academy of Sciences on April 24, 1996.

SHELLEY LAU

Director of Home Affairs

Hong Kong

I ncreasingly, knowledge-based development underpins Hong Kong's economic strategy. It is no secret that Hong Kong is on the brink of a new future as part of China, the largest, most populous nation on earth. Yet, even now, the Hong Kong economy supports not just the six million people in its own territory, but also another four million who are economically linked across the border with China. With the current pace of innovation and change in the world, and with the growing strength of our economic competitors in Southeast Asia, it has become clear that the trade and manufacturing base of the Hong Kong economy must make a leap of technological innovation into the future.

New Initiatives in Science and Technology—and the Link to China

Although perhaps a more recent member of the Asian S&T fan club, Hong Kong has made quick and decisive strides. Indeed, the level and pace of our investments in science and technology are evidence of the confidence we have in the future and the transition to being part of China.

> The level and pace of our investments in science and technology are evidence of the confidence we have in the future and the transition to being part of China.

In 1990, we established the Hong Kong University of Science and Technology to serve as a flagship for academic innovation. We are now in the midst of developing a $2.3 billion science park to serve as an industrial anchor for high-tech innovation and manufacturing. In addition to these investments, the Hong Kong government has put in place a series of policy initiatives to more deeply link technology to economic growth.

The Industry Department of the Hong Kong government has established a number of financing strategies to help local manufacturers upgrade their processes and technology. For example, the Cooperative Applied Research and Development Scheme, also known as CARDS, facilitates research relationships in manufacturing between Hong Kong and China. Indeed, CARDS has attracted the attention of both international investors and scientists from around the world who want to be in on the ground floor of the technological advances that may come from Hong Kong's China transition. In addition, seven overseas investment promotion funds have been established around the world to try to attract increasing levels of technology investment to Hong Kong.

Much of the innovation in Hong Kong's industries is a product of the efforts of the Hong Kong Productivity Council. This statutory body is several decades old and was initiated to upgrade the level of R&D in Hong Kong's industry. Many of Hong Kong's industrialists came from China in the 1950s, and their view of industrial development was very, very traditional. By the 1970s, it became clear that competitiveness in virtually all global markets would require a new emphasis on research and on upgrading manufacturing technology throughout the industrial sector. The Productivity Council was created to take on that task, and it is largely responsible for the level of technological advance in industry that is seen in Hong Kong today.

> **One emerging concern is how to maintain our standards, which are of international quality and substance.**

Maintaining Standards

One concern, however, is emerging as we move toward the transfer of sovereignty in 1997. In Hong Kong we have been very careful to establish world-class standards in the economy, with regard to technology, management, finance, and the like. We are all concerned that the transition take place in such a way that Hong Kong and China mutually benefit from their new relationship, but that the standards underpinning our own industrial infrastructure, our R&D, and our human resources remain at the very highest levels.

The "Basic Law" as a Guide for the S&T Future

One of the key assurances in maintaining these standards is in the language of the Basic Law, which sets out the legal parameters for Hong Kong's new relationship with China. The Basic Law is, in

> **Article 139 of the Basic Law explicitly states that the Hong Kong special administrative region government shall, on its own, formulate policies on science and technology and protect achievements in research, patents, discoveries, and inventions.**

effect, a mini-constitution for Hong Kong. It was promulgated after the joint declaration between China and the United Kingdom was signed. Article 139 of the Basic Law deals with science and technology. That article explicitly states that the Hong Kong special administrative region government shall, on its own, formulate policies on science and technology and protect by law achievements in scientific and technological research, patents, discoveries, and inventions. Article 139 further states that, on its own, Hong Kong shall decide on the scientific and technological standards and specifications applicable in Hong Kong.

The operative phrase is this article is "on its own." Many people fear the transition and feel that China will overwhelm Hong Kong. The Basic Law states categorically that Hong Kong will, on its own, formulate policies specifically related to science-based or knowledge-based development. The law guarantees that, in the future, Hong Kong will have the autonomy, flexibility, and authority to follow its own practices and relate to the international community in its pursuit of knowledge-based development.

The Hong Kong Way

What are the characteristics, or standards, of Hong Kong policy that it will continue to pursue? And is there any reason to expect future policy conflicts? I believe closer examination will reveal little for anyone to be concerned about. Hong Kong has approached the considerable problems involved in promoting science and technology in a very sensible way. First and most importantly, Hong Kong has provided a favorable macroeconomic environment by maintaining a strong economy and optimizing public investment. In S&T decisions, we have taken that economic base and approached investments from the point of view of developing consensus between government and industry about what resources and strategies best link S&T to economic growth. The S&T policy emphasis has always been on investment—in science, in technology, in knowledge.

> **Government plays the role of catalyst. We do not, and will not, play mother hen to industry. We will not interfere in the marketplace.**

With this strategy, government plays, and will continue to play, the role of catalyst and facilitator. We do not, and will not, play mother hen to industry. We will not interfere in the marketplace.

Hong Kong will continue to explore ways to collaborate with China on science and technology development. We have much to share and much in common. Many opportunities will fit quite well into the Hong Kong emphasis on maximizing investment in knowledge. Indeed, much collaboration is currently under way.

The Case of Chinese Medicine

Chinese medicine provides a concrete example of this type of collaboration. Article 138 of the Basic Law states that Hong Kong shall, on its own, develop Chinese medicine in the same way that it develops Western medicine. There is great international interest in the potential of Chinese medicine, and yet there are many unresolved issues regarding its efficacy and its use in relation to Western medicine. We in Hong Kong are working closely with China on applying the methodology of modern science to the study of Chinese medicine. We have worked with universities and medical schools in Beijing, Shanghai, and Nanjing to examine their process

> **Hong Kong is facing the future with optimism and energy. Ten years ago, the dominant emotion was anxiety. Today, it is enthusiasm.**

of medical school instruction, which integrates Chinese medicine with Western medicine. We are quietly developing joint projects.

So, at the project level, it is quite clear the S&T collaboration is both possible and fruitful.

A Concluding Thought

Hong Kong is embracing the future with optimism and energy. Ten years ago, the dominant emotion inside and outside the government was anxiety. Today, anxiety has been replaced with enthusiasm. We recognize that problems will arise. But they can be overcome. Regardless of whatever challenges lie ahead, we will continue to maintain our traditions and values. We face the future with a bit of wisdom from Abraham Lincoln in the back of our minds. He once said, "I'm not bound to win, but I am bound to be true. I'm not bound to succeed, but I am bound to live up to what light I have. I must stand with anybody that stands right, stand with him while he is right and part with him when he goes wrong." ■

The Globe as the Market: The Challenge of Competitiveness for U.S. Industry

Based on remarks delivered at the New York Academy of Sciences on February 23, 1995.

DANIEL BURTON
President, Council on Competitiveness

here the U.S. stands in terms of global competitiveness depends, to some extent, on the indicators chosen to ascertain position and momentum. Traditional economic indicators may tell a very different story than indicators based on global technology use and diffusion. Indeed, the competitiveness debate is in the midst of transition. In the 1980s, the concern was denominated in terms of classic struggles for productivity and quality in manufacturing industries. In the 1990s and into the future, even the definition of economic products will change, and so must the analysis of competitiveness.

Predicting Competitiveness Using Standard Measures

Bright and Sunny

When compared to much of the 1980s, the performance of the U.S. economy relative to Europe and Japan in the early 1990s is a cause for celebration. In 1993, the U.S. standard of living rose more than in any other G-7 country (U.S., U.K., France, Germany, Japan, Italy, Canada). In 1994, the U.S. exported over $600 billion in merchandise, goods, and services, more than any other nation in the world. Productivity also rose by over 4 percent, the largest gain since 1987.

> **The competitiveness debate is shifting from a concern with traditional macroeconomic indicators in an industrial economy to a new focus on technological indicators in a networked economy.**

Moreover, a 1994 Council on Competitiveness survey of the nation's chief executive officers and leading university presidents found similar optimism. Management quality and technology were rated as very good and getting better. The human resources emerging from the university system were also viewed as excellent. Taken together, such standard measures can create a certain complacency about the global competitiveness of U.S. industry.

. . . With Storm Clouds Gathering

But complacency is most definitely unjustified. Although indicators for the most recent year appear vibrant, they are exceptions. The rise in the standard of living in 1994 stands in contrast to the last 20 years, during which the U.S. had the lowest growth rate of any G-7 country.

The same 1994 survey of executives echoed concern over long-

> In terms of savings—the capital for tomorrow's investments—U.S. rates are in the global cellar. America's gross savings rate is half that of Indonesia, Thailand, Hong Kong, or Singapore.

term weaknesses in the U.S. position on two particular measures: savings and education.

Since 1973, the U.S. net national savings rate has hovered at around 2 percent, about one-tenth that of Japan, one-third that of Germany, and less than half that of France and Italy. For six of the past seven years, U.S. investment in plant and equipment was also the lowest of any G-7 country. Comparisons with Asia are striking: The U.S. gross savings rate is about half that of Hong Kong, Indonesia, Singapore, South Korea, Taiwan, or Thailand. Asian countries are accumulating capital at twice the rate of the United States.

The second dominant concern is weaknesses in the nation's K-12 and worker training systems. Changes in the economy and in the nature of the nation's industrial base have not been matched by changes in the technological preparation of the labor force. The traditional manufacturing base in the United States was built around apprenticeships. There was no significant premium on superior primary and secondary education. This is no longer the case, especially as the economy moves to the provision of advanced services. Overall, education and training programs have not kept pace.

Sorting the Near Term From the Long Term

These two very different pictures reveal near-term improvements and long-term problems. Private executives see hard times ahead. Despite a decade of restructuring, layoffs, and downsizing, the executives surveyed, by a margin of two to one, predicted that U.S. industry will continue to face difficult competitiveness hurdles in the future.

After the last five years of focusing internally and managing a difficult reunification process, Germany will return to the international market. It will become an economic power-house in Europe that the world has not seen in half a century. Japan,

> In the 1980s, a lot of the competitiveness debate was about quality. Now, the driving force of com-petitiveness is innovation.

despite current economic weaknesses, will also return to be a strong competitor. The difference is that it may not come back from Japan. Japanese technology, capital, and managers are pervasive throughout Asia, and Japanese competition will come from Taiwan, Indonesia, Malaysia, Singapore, and China.

Innovation Will Drive Long-Term Competitiveness

In the 1980s, much of the competitiveness debate was about quality. American cars and semiconductors were said to be the vital sign of U.S. competitiveness, and they were simply not of a high enough quality to compete in global markets. But U.S. quality has improved. The next hurdle for competitiveness will be innovation.

Innovation implies more than simply focusing on product improvement. It means continuous learning, constant improve-ment, and the creation of entirely new types of enterprise. Critical to the health of innovation is the commitment to research and development. Over the long term, the pace of innovation, and hence the competitiveness of U.S. industry, will depend on invest-ments in research and development.

To ensure those investments, the nation must have a technolo-gy policy. Over the last four decades, the United States has had a defense policy, a space policy, a basic research policy, and a vari-ety of other concerted efforts to address national S&T issues. But there has never been a policy designed to stimulate industrial productivity and industrial technology.

A New Set of Indicators: Technology Within a Networked Economy

The U.S. economy is moving from a traditional industrial structure to a networked structure. Indeed, it is farther along this path than virtually any other economically advanced nation. When U.S. competitiveness is viewed through this new set of optics, a very different picture of the U.S. role in the world emerges. A comparison with Japan is instructive.

> **An economy based on technological networks has different rules of competition from those of an industrial economy**

The number of personal computers per 100 workers in the U.S. is four times that in Japan. In the United States there were 3,900 domestic commercial databases in 1992; in Japan there were 900. In the U.S. there were between two and three million people signed up on the Internet in 1992. In Japan in 1994, there were 39,000. In the U.S., there are about 4.4 local area networks per 100 personal computers. In Japan there are 1.4.

These are not macroeconomic indicators. They do not purport to address economic growth rates or trade balances or productivity. They do not measure traditional economic parameters. But they do show a very different economy, one which is moving very quickly into a series of networked relationships that will drive economic activity in the future.

Moreover, the U.S. role in producing as well as using these technologies is improving. In 1990, a Council on Competitiveness survey showed the U.S. was weak or losing badly in a third of the top 100 technologies deemed to be critical to economic advance over the next decade. In 1994, the same survey showed very different results. In 20 out of the 30 technologies in which the U.S. had lagged, American private industry was growing and regaining competitive ground. America's position in critical technologies has changed from chronic gradual erosion to a gathering strength.

The Role of Telecommunications

The vital sign for this new economy is the telecommunications industry. Telecommunications are the bedrock of the national information infrastructure, which itself will broadly affect the future strength of the U.S. economy.

Five years ago, many observers—both in government and in the private sector—were concerned that directed efforts in France, Singapore, Germany, and Japan were resulting in levels of investment in information infrastructure that would put those economies significantly ahead of the United States. Today the perspective has changed. Instead, the emphasis has shifted from publicly supported telecommunications investment to deregulation. The United Kingdom is now viewed as one of the most advanced in terms of telecommunications investment because it is also one of the most deregulated telecommunications environments in the world.

Whichever policy mode ultimately proves best—directed public programs or widespread deregulation—it is clear that investments in telecommunications will prove key to the economy of tomorrow.

Changing Focus of Private Initiative

Much of the momentum that has been gained in U.S. competitiveness over the last forty years has come from concerted private sector efforts. In the private sector, the emphasis has been on streamlining, cost reduction, efficiency increases, productivity enhancement. All of these initiatives in the industrial economy came as a

> **Alongside a series of very vibrant, positive strengths, the U.S. economy displays two central weaknesses—the quality of primary and secondary education and the level of savings and investment rates.**

response to market forces. The emphasis was on maintaining the nation's industrial leadership in global markets.

The private sector focus has begun to change over the last five years, and will change further into the next century. Private industry is in the process of reinventing R&D. Centralized R&D institutions, akin to the old Bell Labs model, are declining or disappearing. The private sector's appreciation of research has not declined, but the process has changed as the product cycle has become more rapid.

The strength of the United States is innovation, flexibility, entrepreneurship, and openness in higher education. Taken together, these features give the nation an advantage in the global marketplace. American industry in the future will do more than streamline and squeeze more productivity out of existing products and markets. Using innovation with technology as building blocks, and with telecommunications as the pathway, U.S. industry will create new markets for goods and services unanticipated only a decade ago. ■

About the Authors

JOHN F. AHEARNE is Director of the Sigma Xi Center, Adjunct Scholar at Resources for the Future, and Lecturer in Public Policy at Duke University. He was Executive Director of Sigma Xi from 1989 to 1996.

ROBERT A. BELL is Vice President of the Consolidated Edison Company of New York, responsible for Research and Development.

MARK BOHANNON is the Chief Counsel for Technology in the U.S. Department of Commerce, and serves as Counsellor to the Under Secretary for Technology.

DANIEL BURTON is President of the private sector Council on Competitiveness.

JOEL E. COHEN has been Professor of Populations and head of the Laboratory of Populations at The Rockefeller University, New York, since 1975.

KENNETH W. DAM is the Max Pam Professor of American and Foreign Law the University of Chicago Law School.

STEPHEN M. DREZNER, a Senior Vice President of RAND, directs the Critical Technologies Institute (CTI).

ROBERT W. GALVIN is Chairman of the Executive Committee at Motorola, Inc., where he has worked since 1940.

JOHN H. GIBBONS is the Assistant to the President for Science and Technology and Director of the White House Office of Science and Technology Policy.

F. WILLIAM HAWLEY is Director of International Governmental Relations at Citicorp/Citibank.

STEPHEN A. HERZENBERG is Executive Director of the Keystone Research Center. Previously Dr. Herzenberg was a senior analyst at the Office of Technology Assessment, a nonpartisan research agency of the United States Congress.

SHELLEY LAU is Director of Home Affairs, Hong Kong.

WILLIAM J. McDONOUGH is President and CEO of the Federal Reserve Bank of New York.

ROBERT K. MORRIS is Director, International Issues, AMP Incorporated.

IAN MORRISON is the former president of the Institute for the Future, a California-based nonprofit private research and analysis firm

RODNEY W. NICHOLS is President and CEO of the New York Academy of Sciences.

MICHEL OKSENBERG became professor of political science and Senior Fellow, Asia Pacific Research Center, at Stanford University in 1995. Previously, he was president of the East-West Center.

THOMAS R. PICKERING has been Under Secretary of State for Political Affairs since May 1997. At the time of his contribution to this volume he was U.S. ambassador to the Russian Federation.

WILLIAM G. PIETERSEN is Chairman of the Institute for the Future and President of Pietersen Consulting.

ARATI PRABHAKAR is Chief Technology Officer and Senior Vice President. At the time of her contribution to this volume, Dr. Prabhakar was Director of the National Institute of Standards and Technology (NIST).

SUSAN U. RAYMOND is Director of Policy Programs at the New York Academy of Sciences.

J. THOMAS RATCHFORD is Director of the Center for Science, Trade, and Technology Policy and Professor of International Science and Technology Policy at George Mason University.

WALTER I. RICKARD is President and CEO at Listing Services Solutions, Inc. in Lumberton, New Jersey.

G. JON ROUSH is a senior fellow of The Conservation Fund and past president of the Wilderness Society.

MARK SCHAEFER is Deputy Assistant Secretary, Water and Science, at the U.S. Department of the Interior. At the time of his contribution to this volume, he was Assistant Director for Environment in the White House Office of Science and Technology Policy.

WILLIAM J. SPENCER has been the President and Chief Executive Officer of SEMATECH in Austin, Texas since October, 1990.

ROBERT C. STEMPEL is Executive Director of Energy Conversion Devices, Inc., an energy and information company headquartered in Troy, Michigan. He retired from General Motors Corporation as Chairman and Chief Executive Officer in November, 1992.

DONALD STOKES was Class of 1943 University Professor of Politics and Public Affairs in the Woodrow Wilson School at Princeton University. He died in January, 1997.

JAMES A. THOMSON has been RAND's president and chief executive officer since August 1989.

JANE WALES is Assistant to the President for International Security at The Rockefeller Brothers Fund. Formerly, she was Associate Director of the White House Office of Science and Technology Policy.

ROBERT S. WALKER represented Pennsylvania in Congress and in the 104th Congress chaired the Committee on Science. Since January 1997, Mr. Walker has been president of the Wexler Group.

VICTOR WOUK is the principal of Victor Wouk and Associates, a consultancy firm for the electric and hybrid vehicle fields.